D0284333

Bright Wings

Bright Wings

An Illustrated Anthology of
Poems About Birds

Edited by BILLY COLLINS
Paintings by DAVID ALLEN SIBLEY

Columbia University Press NEW YORK

Columbia University Press
Publishers Since 1893
New York   Chichester, West Sussex
Introduction © 2010 Billy Collins
Paintings © David Allen Sibley

Library of Congress Cataloging-in-Publication Data
Bright wings : an illustrated anthology of poems about birds /
edited by Billy Collins / paintings by David Allen Sibley.
p.   cm.
Includes bibliographical references and index.
ISBN 978-0-231-15084-2 (cloth : alk. paper)
1. Birds—Poetry. 2. English poetry. 3. American poetry. I. Collins, Billy.
II. Sibley, David 1961– III. Title.

PR1195.B5B75 2009
821'.008'935—dc22   2009019318

Columbia University Press books are printed on permanent and durable acid-free paper.
This book is printed on paper with recycled content.
Printed in the United States of America

c 10 9 8 7 6 5 4 3 2 1

CONTENTS

. . . .

• • • • • • • • • • • • • • • • • • • •

● ● ● ● ● ● ● ● ● ● ● ● ● ● ● ● ● ● ●

ACKNOWLEDGMENTS

· · · · · · · · · · · · · · · · · · · ·

I would like to thank Tessa Kale, my editor, for her helpful suggestions and her unwavering support, as well as Suzannah Gilman, who assisted me in many aspects of this project and tracked down many of these flighty poems. I am also grateful to all the poets who generously contributed their work and to those who pointed me unselfishly to the poems of others.

Bright Wings

INTRODUCTION

. . . .

Billy Collins

The phenomenon of poems about birds extends from the earliest evidence of writing to a poetry workshop that might have been held last night. To follow the trajectory of these poems is to observe the persistence of the human fascination with these "things with feathers"—Emily Dickinson's image for Hope—that amaze us with their ability both to fly and to sing. But also revealed in such poems are the ways in which attitudes toward birds change as history and culture run their courses. Birds may have always sung their same songs and flown with their same wing beats, but our perceptions of them have vacillated since ancient times. Homer refers to birds as omens as well as carrion-eating reminders of death. In early English poetry, birds can be emblematic (the royal eagle), mythological (the reborn phoenix), or symbolic (the self-wounding pelican as Christ). Birds have their place on the pages of bestiaries, and they even speak in Chaucer's "Parliament of Fowls" and other allegorical dramas. In John Skelton's "Philip Sparrow," dozens of anthropomorphic birds attend the funeral of a sparrow slain by a cat. Birds are prominent in love poetry as messengers and signs of elation. Birds can also signify domesticity in their building of nests and care of fledglings. But it is with the poets of the early nineteenth century that birds achieve their most familiar and powerful poetic status as both symbols of imaginative freedom in their flight and substitutes for the poet in the full-throated ease of their singing.

The most common Romantic reaction to avian sightings is jealousy. The poet envies the spiritual and imaginative liberation suggested by bird flight, and birdsong leaves him with the desire to sing as naturally and as beautifully in his own lyric. "I would be a bird," opens a poem by Robert Burns, "and straight on wings I arise." Such flighty yearnings often fall into the kind of theatrical excess that can give poetry a bad name. This is especially true if the poems are post-Darwinian compositions that lightly step over the inconvenience of the red tooth and claw of the natural world. Stephen Vincent Benét starts a poem with the embarrassing rhyme "Lord, may I be / A sparrow in a tree." "Could, I, too!" wishes F. Grosse. "Could I skim with thee," opines Lucy Aiken in the exclamatory-conditional mood.

Throwing questions at birds, often regarding their mysterious comings and goings, is another habit of poetry in English. "Crane, whence dost thou come?" asks Leo Alishan. If the bird is not being interrogated, it is sometimes being ordered about. "Fly to my birthplace . . . / O swallow, build thy nest," writes C. A. Totochian. "Sing on . . . / Blithe spirit of the morning air," commands Alexander Posey. This kind of bird enthusiasm, which rose to popularity in the nineteenth century and was notoriously practiced by writers with three, even four, names, adds up to a kind of school of wistfulness, marked by the poet's yearning to join a bird in flight or in song and thereby elevate his communion with the creature from the vicarious to the real. Judging by the quantity of poems containing an astonished exclamation at the sight or sound of a bird, a reader of nine-teenth-century poetry might get the impression that poets are incapable of seeing a creature with wings without yelling "Hark!"

But such flighty Romantic yearnings find a balance in more level-eyed descriptive poems informed by actual observation. Most bird poems can be located on a scale that ranges from naturalistic examination to Romantic conjuring of creatures that heighten their elusiveness by remaining invisible. In the first, the eye dominates as John Clare scrutinizes a nest of ink-speckled eggs; in the second, the ear takes over as Keats listens in the dark to the nightingale. Some recent poems about birds may fall into the loose category of "ecopoetry," or they may remain in a state of post-Emersonian idealism regarding nature. Or, aware of the tradition looming behind them, contemporary poets tend to undercut the conventions of ornithological poetry. Anxious about influence, they no longer evoke the song of a lark or the grace of a swan with a straight face and any hope of originality. Still, poets are reluctant to admit that any subject's metaphoric possibilities have been depleted. They simply have to push at the edges of what has been done. Here is a striking example by Ruth Schwartz of how a poet can freshen a tired image; her poem is also proof that poets are people who have trouble saying one thing at a time.

. . . .

The Swan at Edgewater Park

Isn't one of your prissy richpeoples's swans
Wouldn't be at home on some pristine pond
Chooses the whole stinking shoreline, candy wrappers, condoms in its
tidal fringe
Prefers to curve its muscular, slightly grubby neck into the body of a
Great Lake,
Swilling whatever it is swans swill,
Chardonnay of algae with bouquet of crud,
While Clevelanders walk by saying Look at that big duck!
Beauty isn't the point here; of course the swan is beautiful,
But not like Lorie at 16, when
Everything was possible—no
More like Lorie at 27
Smoking away her days off in her dirty kitchen,
Her kid with asthma watching TV,
The boyfriend who doesn't know yet she's gonna
Leave him, washing his car out back—and
He's a runty little guy, and drinks too much, and
It's not his kid anyway, but he loves her, he
Really does, he loves them both—
That's the kind of swan this is.

Besides looking at birds from odd angles, some contemporary poets
exploit the information that newer technologies have made available,
such as photographic studies of bird flight and in-nest behavior. The

obvious way that birds have inspired poets is through their double ability to fly and to sing, often at the same time. But some birds can also amaze us through their habits of work—nest building, mating, parenting, and migrating. Scientific tracking is now revealing some startling facts about their migratory feats. Eamon Grennan's lovely poem, included in this book, bears this moving, explanatory title: "On a 3½ oz. Lesser Yellowlegs, Departed Boston August 28, Shot Martinique September 3." And recently, many jaws dropped at the news that a female Bar-tailed Godwit—a shorebird with a long beak—was recorded by satellite to have flown nonstop more than seven thousand miles from Alaska to New Zealand. Some day her poem might be written, but for now she is known only by her scientific tag: "E7."

Clearly, there are a lot of bird poems flying around out there, and many of them can be found perched on the branches of anthologies. But this assemblage offers a fresh approach. By matching a rich set of poems about birds with illustrations by the illustrious David Allen Sibley, we have provided the reader in one volume with pleasures that are literary, pictorial, and scientific.

Because this gathering did not want merely to echo the work of past anthologizers, many of the obvious choices were passed over. Classics such as Keats's and Coleridge's nightingales, Yeats's swans at Coole, Bryan's waterfowl, Jeffers's hawks, Hopkins's windhover, and Poe's raven have been showcased in so many books of poetry—bird oriented or otherwise—that no editorial regrets were felt at the decision to leave them out. Instead, air time is given to many lesser-known poems, particularly more contemporary ones, in order to give the reader a better chance of being taken by surprise.

. . . .

The usefulness of poetry has often been questioned. "Poetry makes nothing happen," W. H. Auden famously declared in a poem that made nothing happen except to cause people to repeat that line to excess. The aesthete's response to poetry's supposed lack of utility is to point out that uselessness is precisely what distinguishes poetry (and other arts) from the world of practical discourse, whose aims are grossly apparent—the stump speech, the business meeting. Of course, if the meaning of "useful" is extended from how to assemble a piece of outdoor furniture to how to engage our verbal intelligence and uplift the human spirit, then poetry may be said to have a purpose. The case for poetry's purpose, if it still needs to be made, becomes clear if we admit the limits of prose. A subject such as birds may have been covered as extensively as possible in prose, but that does not mean, as with any topic of human interest, that there is nothing left to say. Indeed, the genre of poetry makes its true appearance at the very point along the line of verbal expression where the possibilities of prose have been exhausted. The job of poetry, we might even say, is to make sure that prose is never allowed to have the last word—which is a good enough way to end this little piece of prose and let poetry do the talking.

. . . and with ah! bright wings.

· · · · · · · · · · · · · · · · · · · ·

Gerard Manley Hopkins, "God's Grandeur"

John James Audubon

Some men live for warlike deeds,
Some for women's words.
John James Audubon
Lived to look at birds.

Pretty birds and funny birds,
All our native fowl
From the little cedar waxwing
To the Great Horned Owl.

Let the wind blow hot or cold,
Let it rain or snow,
Everywhere the birds went
Audubon would go.

Scrambling through a wilderness,
Floating down a stream,
All around America
In a feathered dream.

Thirty years of traveling,
Pockets often bare,
(Lucy Bakewell Audubon
Patched them up with care).

Followed grebe and meadowlark,
Saw them sing and splash.
(Lucy Bakewell Audubon
Somehow raised the cash).

Drew them all the way they lived
In their habitats.
(Lucy Bakewell Audubon
Sometimes wondered "Cats?")

Colored them and printed them
In a giant book,
"Birds of North America"—
All the world said, "Look!"

Gave him medals and degrees,
Called him noble names,
—Lucy Bakewell Audubon
Kissed her queer John James.

STEPHEN VINCENT BENÉT

Common Loon

The eerie call of the Common Loon can be heard at night across Canada and the northern United States. It has been variously described as a yodel and a wailing *aarOOOOOaaaa*.

Loons Mating

Their necks and their dark heads lifted into a dawn
Blurred smooth by mist, the loons
Beside each other are swimming slowly
In charmed circles, their bodies stretched under water
Through ripples quivering and sweeping apart
The gray sky now held close by the lake's mercurial threshold
Whose face and underface they share
In wheeling and diving tandem, rising together
To swell their breasts like swans, to go breasting forward
With beaks turned down and in, near shore,
Out of sight behind a windbreak of birch and alder,
And now the haunted uprisen wailing call,
And again, and now the beautiful sane laughter.

DAVID WAGONER

Yellow-nosed Albatross

The endangered Yellow-nosed Albatross can grow up to three feet long. Albatrosses have the largest wingspans of any bird, from more than five to about eleven feet. This species nests on islands in the middle South Atlantic and, at sea, ranges across the South Atlantic.

Drifting Off

The guttersnipe and the albatross
gliding for days without a single wingbeat
were equally beyond me.

I yearned for the gannet's strike,
the unbegrudging concentration
of the heron.

In the camaraderie of rookeries,
in the spiteful vigilance of colonies
I was at home.

I learned to distrust
the allure of the cuckoo
and the gossip of starlings,

kept faith with doughty bullfinches,
levelled my wit too often
to the small-minded wren

and too often caved in
to the pathos of waterhens
and panicky corncrakes.

I gave much credence to stragglers,
overrated the composure of blackbirds
and the folklore of magpies.

But when goldfinch or kingfisher rent
the veil of the usual,
pinions whispered and braced

as I stooped, unwieldy
and brimming,
my spurs at the ready.

Seamus Heaney

Wilson's Storm-Petrel

The abundant Wilson's Storm-Petrel, when in flight, performs an interesting "foot-pattering" dance on the surface of the water while feeding. A pelagic bird, it comes ashore only to breed on the Antarctic coast and nearby islands and is common in the North Atlantic.

Mother Carey's Hen

There are days I don't think about the sea;
      weeks wash by, in fact,
then a shearwater—or some such—flutters by
on the salt flats fanning out in my mind's eye,
reflected there, a shimmering reverie,
        recalling the pact

I once made (and renew today) to hold
      to a higher altitude.
But note the difference between this bird
and me: a slight disruption or harsh word
and I crash, folded seaward, letting cold
        life intrude;

whereas the petrel, mindless of such height,
      scales each watery hill
that rises up, adapting to the shape
of each impediment, each low escape
instinct in it, the scope of its flight
        fitted to its will.

DAS 2015

Brown Pelican

Although a large bird, the Brown Pelican is the smallest of the seven species of pelicans. When hunting for fish, it is capable of twisting in the air while making sudden head-first plunges into the water to catch its prey. It lives on the coasts of the southern United States to northern South America.

The Birds

I'll miss the small birds that come
for the sugar you put out
and the bread crumbs. They've

made the edge of the sea domestic
and, as I am, I welcome that.
Nights my head seemed twisted

with dreams and the sea wash,
I let it all come quiet, waking,
counting familiar thoughts and objects.

Here to rest, like they say, I best
liked walking along the beach
past the town till one reached

the other one, around the corner
of rock and small trees. It was
clear, and often empty, and

peaceful. Those lovely ungainly
pelicans fished there, dropping
like rocks, with grace, from the air,

headfirst, then sat on the water,
letting the pouch of their beaks
grow thin again, then swallowing

whatever they'd caught. The birds,
no matter they're not of our kind,
seem most like us here. I want

to go where they go, in a way, if
a small and common one. I want
to ride that air which makes the sea

seem down there, not the element
in which one thrashes to come up.
I love water, I *love* water—

but I also love air, and fire.

ROBERT CREELEY

Rendings grunts after so much quiet: look:

tide is advancing—billows, mullet leaping toward shore:

also pigfish pinfish herring sheepshead silverside grass and top

minnows prawns:: brown pelicans—Audubon drawn

chestnut cross-hatch iris blue rim reddened yellow tuft:

pistol-shot from wharves: beautiful evolutions

above the leaping shoal:: shot after shot::

made gumbo: salted: smoked: sensible to cold::

muting so profusely not a spot of green's left on the glossy mangrove::

esophagus storing fish: air pockets to cushion the blow:

black banners the drying wings after what is left

of watermire: wearing it: sitting in plainest light.

DAS 2003

Magnificent Frigatebird

The Magnificent Frigatebird—also known as the Man-'o-War Bird and the Frigate Pelican—has the longest wings of any bird relative to its weight. The male has a scarlet throat pouch that, during breeding season, can be inflated like a balloon. It is widespread in the tropical Atlantic.

The Magnificent Frigatebird

They're bullies and the way they feed is gross,
forcing the smaller fishers to disgorge
their catch in flight, then swooping down to snatch
it for themselves. Along the beach they court
in gangs, frenetically, lacking the charm
of strolling balladeers. Absurdly they
all clack their curious bills and flap their wings,
fluttering for the females overhead,
who then fly off with them to strange lagoons.
Honeymoons there are brief because the males,
like feathery Casanovas, soon decamp,
eager for more romance, stranding their mates,
who contemplate the need to nest alone
in what magnificence the marsh affords.

Rapidly cruising or lying on the air there is a bird
    that realizes Rasselas's friend's project
    of wings uniting levity with strength. This
        hell-diver, frigate-bird, hurricane-
bird; unless swift is the proper word
      for him, the storm omen when
    he flies close to the waves, should be seen
      fishing, although oftener
      he appears to prefer

to take, on the wing, from industrious crude-winged species,
    the fish they have caught, and is seldom successless.
    A marvel of grace, no matter how fast his
      victim may fly or how often may
turn. The others with similar ease,
    slowly rising once more,
      move out to the top
      of the circle and stop

and blow back, allowing the wind to reverse their direction—
    unlike the more stalwart swan that can ferry the
    woodcutter's two children home. Make hay; keep
      the shop; I have one sheep; were a less
limber animal's mottoes. This one
    finds sticks for the swan's-down-dress
    of his child to rest upon and would
      not know Gretel from Hänsel.
      As impassioned Handel—

meant for a lawyer and a masculine German domestic
    career—clandestinely studied the harpsichord
    and never was known to have fallen in love,
        the unconfiding frigate-bird hides
in the height and in the majestic
        display of his art. He glides
    a hundred feet or quivers about
        as charred paper behaves—full
        of feints; and an eagle

of vigilance. . . . *Festina lente*. Be gay
    civilly? How so? "If I do well I am blessed
    whether any bless me or not, and if I do
        ill I am cursed." We watch the moon rise
on the Susquehanna. In his way,
        this most romantic bird flies
    to a more mundane place, the mangrove
        swamp to sleep. He wastes the moon.
        But he, and others, soon

rise from the bough and though flying, are able to foil the tired
    moment of danger that lays on heart and lungs the
        weight of the python that crushes to powder.

· · · · · · · · · · · · · · · · · · ·

DAS 2008

Blue-footed Booby

The word "booby" comes from the Spanish *bobo*, which means "clown" or
"fool." Clumsy-looking on land, the Blue-footed Booby is graceful in water.
Its permanently closed nostrils are an adaptation for diving that allows it to feed
on school fish and squid while still underwater. To court the female, the male flaunts
his bright-blue feet in a dance. It lives on the Galápagos Islands and other Pacific
tropical and subtropical islands.

The Blue Booby

The blue booby lives
on the bare rocks
of Galapagos
and fears nothing.
It is a simple life:
they live on fish,
and there are few predators.
Also, the males do not
make fools of themselves
chasing after the young
ladies. Rather,
they gather the blue
objects of the world
and construct from them

a nest—an occasional
Gaulois package,
a string of beads,
a piece of cloth from
a sailor's suit. This
replaces the need for
dazzling plumage;
in fact, in the past
fifty million years
the male has grown
considerably duller,
nor can he sing well.
The female, though,

asks little of him—
the blue satisfies her
completely, has
a magical effect
on her. When she returns
from her day of
gossip and shopping,
she sees he has found her
a new shred of blue foil:
for this she rewards him
with her dark body,
the stars turn slowly
in the blue foil beside them
like the eyes of a mild savior.

DAS 2007

Great Blue Heron

The Great Blue Heron can grow to almost five feet tall, with a wingspan of more than six feet. It is both the largest and the most widespread heron in North America. A white form can be found in southern Florida and the Caribbean.

Solitary, silent at the brown burn's edge,
   Bent above the ripple where the shy trout run,
He but sees the wan wave lapping on the sedge —
   I can see the bit-bars flashing in the sun.

High and swift above him rush the startled teal,
   Grey and close about him folds the mother-mist,
He but sees the round hill rising like a wheel —
   I can see a horseman with hawk upon his wrist.

Brown below the heather runs the ripple on his feet,
   Low among the shadows there are shadows slipping through,
He but sees the moor-trout mingling as they meet —
   I can see the goshawk stooping from the blue.

Now he hears a footstep; wakes a sleeping power;
   Wide-winged and wonderful sails away, and slow. —
I can see a tall knight 'neath a lady's bower
   Riding with a shorn plume at his saddle-bow.

WILL H. OGILVIE

All winter
the blue heron
slept among the horses.
I do not know
the custom of herons,
do not know
if the solitary habit
is their way,
or if he listened for
some missing one—
not knowing even
that was what he did—
in the blowing
sounds in the dark.
I know that
hope is the hardest
love we carry.
He slept
with his long neck
folded, like a letter
put away.

JANE HIRSHFIELD

Mute Swan

The Mute Swan is not actually mute but gives a variety of calls, ranging from a clear, bugling sound like that of the Tundra Swan to the hisses and snorts made by all swans. It is a native of northern and central Eurasia.

On the Sight of Swans in Kensington Gardens

Queen-bird, that sittest on thy shining nest
And thy young cygnets without sorrow hatchest,
And thou, thou other royal bird, that watchest
Lest the white mother wandering feet molest:
Shrined are your offspring in a crystal cradle,
Brighter than Helen's ere she yet had burst
Her shelly prison. They shall be born at first
Strong, active, graceful, perfect, swan-like, able
To tread the land or waters with security,
Unlike poor human births, conceived in sin,
In grief brought forth, both outwardly and in
Confessing weakness, error, and impurity.
Did heavenly creatures own succession's line,
The births of heaven like to yours would shine.

CHARLES LAMB

On the Marriage of Friends

So you have chosen the way of the swan;
the way, perhaps, that is not natural
to everyone, but I will not harp on
about heron, bluebird or dotterel,

nor how the male flycatcher pairs
with two females, keeping a mile between,
so neither cops how the other shares
the same philandering gentleman.

Did you know the life-coupling way
of the swan is also that of the crow?
And there'll be crow-black days
you'll caw at each other with blind gusto.

But there'll be times when you'll sing
the duet of the black-collared barbet,
with the first part of the song sung
by one and the second by the mate.

We wish you now many such duet days
and sing for you like the red-eyed vireo
who sings nonstop through the summer blaze
on this day you take the way of swan & crow.

GREG DELANTY

Swan and Shadow

```
                 Dusk
             Above the
        water   hang   the
                 loud
                 flies
                 Here
                 O so
                 gray
                 then
             What                A pale signal will appear
             When               Soon before its shadow fades
             Where         Here in this pool of opened eye
             In us     No Upon us As at the very edges
          of where we take shape in the dark air
            this object bares its image awakening
              ripples of recognition that will
                brush darkness up into light
even after this bird this hour both drift by atop the perfect sad instant now
              already passing out of sight
            toward yet-untroubled reflection
            this image bears its object darkening
          into memorial shades Scattered bits of
          light       No of water Or something across
          water          Breaking up No Being regathered
          soon          Yet by then a swan will have
          gone                Yes out of mind into what
             vast
             pale
             hush
             of a
             place
             past
        sudden dark as
           if a swan
             sang
```

JOHN HOLLANDER

Snow Goose

The Snow Goose breeds in scattered colonies in Alaska, Arctic Canada, Greenland, and northeastern Siberia, and winters primarily in central California, on the western Gulf Coast, and on the Mid-Atlantic Coast. Vagrants, though, can be found in Great Britain and Ireland, where they are seen among flocks of Brent and Barnacle geese.

Wild Geese

You do not have to be good.
You do not have to walk on your knees
for a hundred miles through the desert, repenting.
You only have to let the soft animal of your body
    love what it loves.
Tell me about despair, yours, and I will tell you mine.
Meanwhile the world goes on.
Meanwhile the sun and the clear pebbles of the rain
are moving across the landscapes,
over the prairies and the deep trees,
the mountains and the rivers.
Meanwhile the wild geese, high in the clean blue air,
are heading home again.
Whoever you are, no matter how lonely,
the world offers itself to your imagination,
calls to you like the wild geese, harsh and exciting—
over and over announcing your place
in the family of things.

MARY OLIVER

Blue-winged Teal

The powder-blue wing patch of the Blue-winged Teal is revealed in flight. After the Mallard, it is the most abundant duck in North America. It migrates over long distances: one teal banded in Alberta was shot in Venezuela a month later.

The Teal

The teal wears such a look as if
It had gazed into the water's depths.

Translated by Asatarō Miyamori

From Woody's Restaurant, Middlebury

Today, noon, a young macho friendly waiter and three diners,
    business types —two males, one female—
are in a quandary about the name of the duck paddling
    Otter Creek,
the duck being brown, but too large to be a female mallard.
    They really
want to know, and I'm the human-watcher behind the nook
    of my table,
camouflaged by my stillness and nonchalant plumage.
    They really want to know.
This sighting I record in the back of my *Field Guide to People*.

GREG DELANTY

Turkey Vulture

A carrion feeder, the Turkey Vulture finds its food by smell. It is usually silent, although it has been known to softly hiss, to cluck, and to whine. Its range extends throughout much of North, Central, and South America.

Still, Citizen Sparrow

Still, citizen sparrow, this vulture which you call
Unnatural, let him but lumber again to air
Over the rotten office, let him bear
The carrion ballast up, and at the tall

Tip of the sky lie cruising. Then you'll see
That no more beautiful bird is in heaven's height,
No wider more placid wings, no watchfuller flight;
He shoulders nature there, the frightfully free,

The naked-headed one. Pardon him, you
Who dart in the orchard aisles, for it is he
Devours death, mocks mutability,
Has heart to make an end, keeps nature new.

Thinking of Noah, childheart, try to forget
How for so many bedlam hours his saw
Soured the song of birds with its wheezy gnaw,
And the slam of his hammer all the day beset

The people's ears. Forget that he could bear
To see the towns like coral under the keel,
And the fields so dismal deep. Try rather to feel
How high and weary it was, on the waters where

He rocked his only world, and everyone's.
Forgive the hero, you who would have died
Gladly with all you knew; he rode that tide
To Ararat; all men are Noah's sons.

RICHARD WILBUR

Northern Goshawk

The widespread Northern Goshawk inhabits the temperate forests of the Northern Hemisphere. An unusually persistent hunter, it sometimes chases prey for up to an hour, often maneuvering through vegetation on its short, broad wings. Attila the Hun had an image of the Northern Goshawk on his helmet.

To a Farmer Who Hung
Five Hawks on His Barbed Wire

They saw you behind your muzzle much more clearly
Than you saw them as you fired at the sky.
You meant almost nothing. Their eyes were turning
To more important creatures hiding
In the grass or pecking and strutting in the open.
The hawks didn't share your nearsighted anger
But soared for the sake of their more ancient hunger
And died for it, to become the emblem
Of your estate, your bloody coat-of-arms.

If fox and raccoon keep out, your chickens may spend
Fat lives at peace before they lose
Their appetites, later on, to satisfy yours.
You've had strange appetites now and then,
Haven't you. Funny quickenings of the heart.
Impulses not quite mentionable
To the wife or yourself. Even some odd dreams.
Remember that scary one about flying?
You woke and thanked the dawn you were heavy again.

Tonight, I aim this dream straight at your skull
While you nestle it against soft feathers:
You hover over the earth, its judge and master,
Alert, alive, alone in the wind
With your terrible mercy. Your breastbone shatters
Suddenly, and you fall, flapping,
Your claws clutching at nothing crookedly
End over end, and thump to the ground.
You lie there, waiting, dying little by little.

You rise and go on dying a little longer,
No longer your heavy self in the morning
But light, still lighter long into the evening
And long into the night and falling
Again little by little across the weather,
Ruffled by sunlight, frozen and thawed
And rained away, falling against the grass
Little by little, lightly and softly,
More quietly than the breath of a deer mouse.

DAVID WAGONER

Evening Hawk

From plane of light to plane, wings dipping through
Geometries and orchids that the sunset builds,
Out of the peak's black angularity of shadow, riding
The last tumultuous avalanche of
Light above pines and the guttural gorge,
The hawk comes.

　　　　His wing
Scythes down another day, his motion
Is that of the honed steel-edge, we hear
The crashless fall of stalks of Time.

The head of each stalk is heavy with the gold of our error.

Look! Look! he is climbing the last light
Who knows neither Time nor error, and under
Whose eye, unforgiving, the world, unforgiven, swings
Into shadow.

　　　　Long now,
The last thrush is still, the last bat
Now cruises in his sharp hieroglyphics. His wisdom
Is ancient, too, and immense. The star
Is steady, like Plato, over the mountain.

If there were no wind we might, we think, hear
The earth grind on its axis, or history
Drip in darkness like a leaking pipe in the cellar.

You and I Saw Hawks Exchanging the Prey

They did the deed of darkness
In their own mid-light.

He plucked a gray field mouse
Suddenly in the wind.

The small dead fly alive
Helplessly in his beak,

His cold pride, helpless.
All she receives is life.

They are terrified. They touch.
Life is too much.

She flies away sorrowing.
Sorrowing, she goes alone.

Then her small falcon, gone.
Will not rise here again.

Smaller than she, he goes
Claw beneath claw beneath
Needles and leaning boughs,

While she, the lovelier
Of these brief differing two,
Floats away sorrowing,

Tall as my love for you,

And almost lonelier.

Delighted in the delighting,
I love you in mid-air,
I love myself the ground.

The great wings sing nothing
Lightly. Lightly fall.

JAMES WRIGHT

from The Parliament of Fowls

There might men the royal eagle find,
That with his sharp look pierceth the sun,
And other eagles of a lower kind,
Of which the clerkes well devise can.
There was the tyrant with his feathers dun
And grey, I mean the goshawk, that doth pine
To briddes for his outrageous ravyne.

DAS 2003

Peregrine Falcon

The powerful and spectacularly fast Peregrine Falcon was once widespread throughout the world and was used in falconry in Europe, Asia, and the Middle East. It was virtually exterminated from eastern North America in the mid-twentieth century because of pesticide poisoning, but a recovery effort has made the raptor a regular sight in a variety of habitats—including large cities. It still is found on every continent except Antarctica.

Peregrine Falcon, New York City

On the 65th floor where he wrote
Advertising copy, joking about
The erotic thrall of words that had
No purpose other than to make
Far too many buy far too much,
He stood one afternoon face to face
With a falcon that veered on the blade
Of its wings and plummeted, then
Swerved to a halt, wings hovering.

An office of computers clicked
Behind him. Below, the silence
Of the miniature lunch time crowds
And toy-like taxis drifting without
Resolve to the will of others.
This bird's been brought in, he thought,
To clean up the city's dirty problems
Of too many pigeons. It's a hired beak.

Still he remained at the tinted glass
Windows, watching as the falcon
Gave with such purpose its self
To the air that carried it, its sheer falls
Breaking the mirrored self-reflections
Of glass office towers. He chided
Himself: this is how the gods come
To deliver a message or a taunt,
And, for a moment, the falcon
Seemed to wait for his response,
The air articulate with a kind of
Wonder and terror. Then it was gone.

He waited at the glass until he felt
The diminishment of whatever
Had unsettled him. And though
The thin edge of the falcon's wings
Had opened the slightest fissure in him
And he'd wandered far in thought,
He already felt himself turning back
To words for an ad, the falcon's power
Surely a fit emblem for something.

ROBERT CORDING

I

Of the oil gland . . . Of the down . . .
    Of the numbers and arrangement
of feathers in the wing . . . I have seen
        on the plains of Apulia

how the birds in earliest spring were weak
    and scarcely able to fly.
Of the avian nostrils and mandibles . . . Of
        the regular sequence of molt . . .

Aristotle, apt to credit hearsay where
    experiment alone
can be relied upon, was wrong about
        the migrant column. Concerning

the methods of capture . . . the jesses . . .
    The swivel, the hood, the falcon's bell . . .

2

The finest of them—here I mean
    for swiftness, strength,

audacity, and stamina—are brooded
    on the Hyperborean cliffs (an island
chiefly made of ice). And I
       am told but have not ascertained

the farther from the sea they nest,
    the nobler will be the offspring.

3

Triangular needles are not to be used.
    The room

to be darkened, the bird
    held close in the hands of the assistant,
linen thread. By no means pierce
      the *membrana nictitans*, lying between

the eyeball and the outermost
    tissue, nor place the suture, lest it tear,
too near the edge. To seel,
      from *cilium*, lower lid,

which makes her more compliant to the falconer's
    will but also (I have
seen this in the lesser birds as well) more bold
      in flight. The senses

to be trained in isolation: taste,
    then touch, then hearing (so
the bars of a song she will evermore link to
        food), and then the sight restored,

in order that the falcon may
    be partly weaned or disengaged from that
which comes by nature.
        The falconer's purse or

carneria, owing
    to the meat it holds . . .
The carrier's arm . . . the gauntlet . . . the horse . . .
        They greatly dislike the human face.

4

If you ask why the train is made of a hare,
    you must know no other flight
more resembles
        the flight at a crane than that

the falcon learns in pursuit of a hare
    nor is more beautiful.
Make-falcon: meaning
        the one who is willing

· ·

to fly in a cast with another less
    expert (the seasons
best suited . . . the weather . . . the hours . . .)
        and by example teach.

5

The removal of dogs, which praise
    will better effect than will the harshest
threats, from the prey. Their reward.
        You must open

the breast and extract the organ that moves
    by itself, which is to say, the heart,
and let the falcon feed.
        The sultan

has sent me a fine machine combining
    the motions of sun and moon,
and Giacomo makes a poem of fourteen
        lines. The music is very good,

I think. (Of those who refuse to come to the lure . . . Of
    shirkers . . . Of bating . . .)
But give me the falcon for art.

LINDA GREGERSON

Bald Eagle

This national emblem of the United States was close to extinction in the contiguous forty-eight states because of DDT poisoning. Protection and reintroduction programs led to its recovery, and the Bald Eagle was reclassified as threatened, rather than endangered, in 1995. Unlike the Golden Eagle—which also lives in Europe, Asia, and North Africa—the Bald Eagle is unique to North America.

The Dalliance of the Eagles

Skirting the river road, (my forenoon walk, my rest,)
Skyward in air a sudden muffled sound, the dalliance of the eagles,
The rushing amorous contact high in space together,
The clinching interlocking claws, a living, fierce, gyrating wheel,
Four beating wings, two beaks, a swirling mass tight grappling,
In tumbling turning clustering loops, straight downward falling,
Till o'er the river pois'd, the twain yet one, a moment's lull,

A motionless still balance in the air, then parting, talons loosing,
Upward again on slow-firm pinions slanting, their separate diverse flight,
She hers, he his, pursuing.

Ring-necked Pheasant

A native of Asia and widely introduced into North America from Eurasia in the mid-nineteenth century, the Ring-necked Pheasant gives a low clucking that sounds much like that of a chicken.

Pheasant

You said you would kill it this morning.
Do not kill it. It startles me still,
The jut of that odd, dark head, pacing

Through the uncut grass on the elm's hill.
It is something to own a pheasant,
Or just to be visited at all.

I am not mystical: it isn't
As if I thought it had a spirit.
It is simply in its element.

That gives it a kingliness, a right.
The print of its big foot last winter,
The tail track, on the snow in our court—

The wonder of it, in that pallor,
Through crosshatch of sparrow and starling.
Is it its rareness, then? It is rare.

But a dozen would be worth having,
A hundred, on that hill—green and red,
Crossing and recrossing: a fine thing!

It is such a good shape, so vivid.
It's a little cornucopia.
It unclaps, brown as a leaf, and loud,

Settles in the elm, and is easy.
It was sunning in the narcissi.
I trespass stupidly. Let be, let be.

Sandhill Crane

The Sandhill Crane has the longest fossil record of any extant bird and so is among the oldest living birds. One fossil places cranes in present-day Nebraska 9 million years ago. Its range is from northeastern Siberia throughout most of North America.

A Bird at the Leather Mill

The crane stood in the center of the floor
of the mill, lost and tentative. Its bill
looked like a fancy awl with a down handle.
It wore its wings as though they were a shawl
thrown on an idiot. At first the men
imagined that a person had strolled in
like a green salesman or a debutant.
And when the crane walked toward the loading dock,
the men on tiptoe prowled with laundry bags
to grab and hold it like a secret hope
harbored in exile. Later on, at lunch,
they took turns, each explaining what he'd do
if it came back. They bragged, or chaffed, aware
the thing was lost, but never saying so.

American Coot

A common waterbird, the American Coot is not a duck, but a rail. Its white bill is triangular like a chicken's, rather than flat like a duck's, and it does not have webbed feet, but lobes on the side of each toe. It is found throughout North and Central America.

Oh Coot! oh bold, adventurous Coot.
 I pray thee tell to me,
The perils of that stormy time,
 That bore thee to the sea!

I saw thee on the river fair,
 Within thy sedgy screen;
Around thee grew the bulrush tall,
 And reeds so strong and green.

The kingfisher came back again
 To view thy fairy place;
The stately swan sailed statelier by,
 As if thy home to grace.

But soon the mountain flood came down,
 And bowed the bulrush strong;
And far above those tall green reeds,
 The waters poured along.

"And where is she, the Water-Coot,"
 I cried, "that creature good?"
But then I saw thee in thine ark,
 Regardless of the flood.

Amid the foaming waves thou sat'st.
    And steered thy little boat;
Thy nest of rush and water-reed
    So bravely set afloat.

And on it went, and safely on
    That wild and stormy tide;
And there thou sat'st, a mother bird,
    Thy young ones at thy side.

Oh Coot! oh bold, adventurous Coot!
    I pray thee tell to me,
The perils of that stormy voyage
    That bore thee to the sea!

Hadst thou no fear, as night came down
    Upon thy watery way,
Of enemies, and dangers dire
    That round about thee lay?

Didst thou not see the falcon grim
    Swoop down as thou passed by?
And 'mong the waving water flags
    The lurking otter lie?

The eagle's scream came wildly near,
    Yet, caused it no alarm?
Nor man, who seeing thee, weak thing,
    Did strive to do thee harm?

And down the foaming waterfall,
    As thou was borne along,
Hadst thou no dread? Oh daring bird,
    Thou hadst a spirit strong!

Yes, thou hadst fear. But He who sees
    The sparrows when they fall,
He saw thee, bird, and gave thee strength
    To brave thy perils all.

He kept thy little bark afloat;
    He watched o'er thine and thee;
And safely through the foaming flood
    Hath brought thee to the sea.

MARY HOWITT

DAS 2002

Mountain Plover

The nesting habitat of the Ringed Plover in the poem is open ground with little or no plant growth. This Mountain Plover, a larger species without the black mask and neckband of its primarily Eurasian cousin, is misnamed: a native of the short-grass prairie, it lives exclusively in the prairie country of North America.

Plover

If, while pitching my tent
west of Lochan na h-Earba,
I'd trodden on the nest
of the ringed plover,

I'd have walked the moor
of brittle grass and tormentil
barefoot, in the summer hail
—distant from that

small bird tilting by the water,
who finds everything
she needs there—shingle
for her four eggs; food, and company.

DAS 2002

American Golden-Plover

The American Golden-Plover prefers pastures to mudflats. It makes one of the longest migratory journeys of any shorebird, from the tundra of Alaska and Canada, where it breeds, to the grasslands of central and southern South America, where it winters.

Shorebird-Watching

To more than give names
to these random arrivals—
teeterings and dawdlings
of dunlin and turnstone,
blackbellied or golden
plover, all bound for

what may be construed as
a kind of avian Althing
out on the Thingstead,
the unroofed synagogue
of the tundra—is already
to have begun to go wrong.

What calculus, what
tuning, what unparsed
telemetry within the
retina, what overdrive
of hunger for the nightlong
daylight of the arctic,

are we voyeurs of? Our
bearings gone, we fumble
a welter of appearance,
of seasonal plumages
that go dim in winter:
these bright backs'

tweeded saffron, dark
underparts the relic
of what sibylline
descents, what harrowings?
Idiot savants, we've
brought into focus

such constellations,
such gamuts of
errantry, the very
terms we're condemned
to try to think in
turn into a trespass.

But Adam, drawn toward
that dark underside,
its mesmerizing
circumstantial thumbprint,
would already have
been aware of this.

DAS 2008

Lesser Yellowlegs

As the poem indicates, the Lesser Yellowlegs travels long distances, migrating
from its breeding grounds in the boreal forests of Alaska and Canada to salt- and
freshwater habitats from the southern United States to southern South America.
It is a regular vagrant to western Europe, with the occasional bird wintering
in Great Britain and Ireland.

On a 3 oz. Lesser Yellowlegs, Departed Boston August 28, Shot Martinique September 3

Little brother, would I could
Make it so far, the whole globe
Curling to the quick of your wing.

You leave our minds lagging
With no word for this gallant
Fly-by-night, blind flight.

But ah, the shot: you clot
In a cloud of feathers, drop
Dead in a nest of text-books.

Now seasons migrate without you
Flying south. At the gunman's door
The sea-grapes plump and darken.

EAMON GRENNAN

Buff-breasted Sandpiper

The Buff-breasted Sandpiper migrates mostly through central North America to southern South America, although flocks of this small shorebird commonly are seen in Great Britain and Ireland. Close to extinction in the 1920s, the species recovered but its numbers may be declining again.

Sandpiper

The roaring alongside he takes for granted,
and that every so often the world is bound to shake.
He runs, he runs to the south, finical, awkward,
in a state of controlled panic, a student of Blake.

The beach hisses like fat. On his left, a sheet
of interrupting water comes and goes
and glazes over his dark and brittle feet.
He runs, he runs straight through it, watching his toes.

—Watching, rather, the spaces of sand between them
where (no detail too small) the Atlantic drains
rapidly backwards and downwards. As he runs,
he stares at the dragging grains.

The world is a mist. And then the world is
minute and vast and clear. The tide
is higher or lower. He couldn't tell you which.
His beak is focussed; he is preoccupied,

looking for something, something, something.
Poor bird, he is obsessed!
The millions of grains are black, white, tan, and gray
mixed with quartz grains, rose and amethyst.

ELIZABETH BISHOP

DAS 2002

Bonaparte's Gull

The Bonaparte's Gull is the only gull that regularly nests in trees. Its name honors the ornithologist Charles Lucien Bonaparte, a nephew of Napoleon who was a member of the Academy of Natural Sciences of Philadelphia in the 1820s. The breeding plumage of both the male and the female is distinguished by a striking black head. Common throughout North America, it is a rare vagrant to western Europe.

The Ballet of the Fifth Year

Where the sea gulls sleep or indeed where they fly
Is a place of different traffic. Although I
Consider the fishing bay (where I see them dip and curve
And purely glide) a place that weakens the nerve
Of will, and closes my eyes, as they should not be
(They should burn like the street-light all night quietly,
So that whatever is present will be known to me),
Nevertheless the gulls and the imagination
Of where they sleep, which comes to creation
In strict shape and color, from their dallying
Their wings slowly, and suddenly rallying
Over, up, down the arabesque of descent,
Is an old act enacted, my fabulous intent
When I skated, afraid of policemen, five years old,
In the winter sunset, sorrowful and cold,
Hardly attained to thought, but old enough to know
Such grace, so self-contained, was the best escape to know.

DELMORE SCHWARTZ

A gull, up close,
looks surprisingly stuffed.
His fluffy chest seems filled
with an inexpensive taxidermist's material
rather lumpily inserted. The legs,

unbent, are childish crayon strokes—
too simple to be workable.
And even the feather markings,
whose intricate symmetry is the usual glory of birds,
are in the gull slovenly,
as if God makes too many
to make them very well.

Are they intelligent?
We imagine so, because they are ugly.
The sardonic one-eyed profile, slightly cross,
the narrow, ectomorphic head, badly combed,
the wide and nervous and well-muscled rump
all suggest deskwork: shipping rates
by day, Schopenhauer
by night, and endless coffee.

At that hour on the beach
when the flies begin biting in the renewed coolness
and the backsliding skin of the after-surf
reflects a pink shimmer before being blotted,
the gulls stand around in the dimpled sand
like those melancholy European crowds
that gather in cobbled public square in the wake
of assassinations and invasions,
heads cocked to hear the latest radio reports.

It is also this hour when plump young couples
walk down to the water, bumping together,
and stand thigh-deep in the rhythmic glass.
Then they walk back toward the car,
tugging as if at a secret between them,
but which neither quite knows—
walk capricious paths through the scattering gulls,
as in some mythologies
beautiful gods stroll unconcerned
among our mortal apprehensions.

JOHN UPDIKE

DAS 2005

Red-legged Kittiwake

The Red-legged Kittiwake may be the least likely gull to be encountered because
it lives on the Bering Sea, nesting on only four island cliffs and then disappearing
out at sea during the winter. The gull in the poem is the Black-legged Kittiwake,
a close relative that is known in Europe as the Kittiwake.

With blistered heels and bones that ache,
Marching through pitchy ways and blind,
The miry track is hard to make;
Yet, ever hovering in my mind,
Above red crags a kittiwake
Hangs motionless against the wind—

Grey-winged, white-breasted and black-eyed,
Against red crags of porphyry
That pillar from a sapphire tide
A sapphire sky. . . . Indifferently
The raw lad limping at my side
Blasphemes his boots, the world, and me. . . .

Still keen, unwavering and alert,
Within my aching empty mind
The bright bird hovers—and the dirt
Of bottomless black ways and blind,
And all the hundred things that hurt
Past healing, seem to drop behind.

WILFRID WILSON GIBSON

Eurasian Collared Dove

The Eurasian Collared Dove, originally from the Middle East, spread to south-eastern Europe in the seventeenth century and across the rest of Europe in the twentieth. Introduced into the Caribbean in the 1970s, the dove reached Florida in the early 1980s. It can be seen occasionally as far north as British Columbia and occupies an ecological niche between the Mourning Dove, which it resembles, and the Rock Pigeon.

After Reading Peterson's Guide

I used to call them
Morning Doves, those birds
with breasts the rosy color
of dawn who coo us awake
as if to say love . . .
love . . . in the morning.

But when the book said
Mourning Doves instead
I noticed their ash-gray feathers,
like shadows
on the underside
of love.

When the Dark Angel comes
let him fold us in wings
as soft as these birds',
though the speckled egg
hidden deep in his nest
is death.

LINDA PASTAN

Red-crowned Parrot

The Red-crowned Parrot is an endangered Amazon parrot native to northeastern Mexico. Feral parrots have appeared in urban areas in southern California, the Rio Grande Valley of Texas, and southern Florida.

She can cry his name from today to tomorrow.
She can Charlie him this, cracker him that, there
in the topmost he hangs like
a Christmas ornament,
his tail
a cascade of emeralds and limes.

The child is heartsick. She has taped messages
to the mailboxes, the names
he responds to, his favorite seeds.
At twilight she calls and calls.

Oh, Charlie, you went everywhere with her,
to the post office and the mall, to the women's
room at the Marriott where you perched
on the stall, good-natured, patient.

And didn't you love to take her thumb
in your golden beak
and, squeezing tenderly, shriek and shriek
as if your own gentleness
were killing you?

You were her darling, her cinnamon stick, her pedagogue.
You knew her secret names
in Persian and ancient Greek. At the beach
you had your own chair and umbrella.
Oh, pampered bird. The neighbors sympathize. But what's
love compared with wild red fruit, a big
gold moon, and an evening that smells of paradise?

If she were older, she'd join the other
sad girls for drinks, she'd lick
the salt from her tequila glass and say something wise
she'd heard said a hundred times before.
Love is a cage she's glad to be free of.

Oh, Charlie, you were her pope and popinjay, her
gaudy, her flambeau, her magnificat.
You were the postcard
each morning delivered to her room, her all-day sunset.

In the topmost fronds you squall and squawk
to the other flashy runaways,
Say paradise! No dice, no dice.

The Contrast; The Parrot and the Wren

I

Within her gilded cage confined,
I saw a dazzling Belle,
A Parrot of that famous kind
Whose name is NON-PAREIL.

Like beads of glossy jet her eyes;
And, smoothed by Nature's skill,
With pearl or gleaming agate vies
Her finely-curvèd bill.

Her plumy mantle's living hues
In mass opposed to mass,
Outshine the splendour that imbues
The robes of pictured glass.

And, sooth to say, an apter Mate
Did never tempt the choice
Of feathered Thing most delicate
In figure and in voice.

But, exiled from Australian bowers,
And singleness her lot,
She trills her song with tutored powers,
Or mocks each casual note.

No more of pity for regrets
With which she may have striven!
Now but in wantonness she frets,
Or spite, if cause be given;

Arch, volatile, a sportive bird
By social glee inspired;
Ambitious to be seen or heard,
And pleased to be admired!

II

This moss-lined shed, green, soft, and dry,
Harbours a self-contented Wren,
Not shunning man's abode, though shy,
Almost as thought itself, of human ken.

Strange places, coverts unendeared,
She never tried; the very nest
In which this Child of Spring was reared
Is warmed, through winter, by her feathery breast.

To the bleak winds she sometimes gives
A slender unexpected strain;
Proof that the hermitess still lives,
Though she appear not, and be sought in vain.

Say, Dora! tell me, by yon placid moon,
If called to choose between the favoured pair,
Which you would be,—the bird of the saloon,
By lady-fingers tended with nice care,
Caressed, applauded, upon dainties fed,
Or Nature's DARKLING of this mossy shed?

WILLIAM WORDSWORTH

DAS 2008

Great Horned Owl

The only animal that regularly eats skunks, the Great Horned Owl also preys on birds, including other owls, nestling Ospreys, and adult and nestling American Crows. Flocks of crows congregate from long distances to mob the owls, sometimes cawing at them for hours. A nonmigratory bird, it has an extensive range—almost all of North America, through Central America, and into South America.

The Great Horned Owl

One morning the Grand Seigneur
Is so good as to appear.
He sits in a scrawny little tree
In my backyard.

When I say his name aloud,
He turns his head
And looks at me
In utter disbelief.

I show him my belt,
How I had to
Tighten it lately
To the final hole.

He ruffles his feathers,
Studies the empty woodshed,
The old red Chevy on blocks.
Alas! He's got to be going.

A Barred Owl

The warping night air having brought the boom
Of an owl's voice into her darkened room,
We tell the wakened child that all she heard
Was an odd question from a forest bird,
Asking of us, if rightly listened to,
"Who cooks for you?" and then "Who cooks for you?"

Words, which can make our terrors bravely clear,
Can also thus domesticate a fear,
And send a small child back to sleep at night
Not listening for the sound of stealthy flight
Or dreaming of some small thing in a claw
Borne up to some dark branch and eaten raw.

RICHARD WILBUR

To a Captive Owl

I should be dumb before thee, feathered sage!
    And gaze upon thy phiz with solemn awe,
But for a most audacious wish to gauge
    The hoarded wisdom of thy learned craw.

Art thou, grave bird! so wondrous wise indeed?
    Speak freely, without fear of jest or gibe—
What is thy moral and religious creed?
    And what the metaphysics of thy tribe?

A Poet, curious in birds and brutes,
    I do not question thee in idle play;
What is thy station? What are thy pursuits?
    Doubtless thou hast thy pleasures—what are *they*?

Or is't thy wont to muse and mouse at once,
    Entice thy prey with airs of meditation,
And with the unvarying habits of a dunce,
    To dine in solemn depths of contemplation?

There may be much—the world at least says so—
    Behind that ponderous brow and thoughtful gaze;
Yet such a great philosopher should know,
    It is by no means wise to think always.

And, Bird, despite thy meditative air,
    I hold thy stock of wit but paltry pelf—
Thou show'st that same grave aspect everywhere,
    And wouldst look thoughtful, stuffed, upon a shelf.

I grieve to be so plain, renowned Bird—
    Thy fame's a flam, and thou an empty fowl;
And what is more, upon a Poet's word
    I'd say as much, wert thou Minerva's owl.

So doff th' imposture of those heavy brows;
    They do not serve to hide thy instincts base—
And if thou must be sometimes munching *mouse*,
    Munch it, O Owl! with less profound a face.

HENRY TIMROD

An Owl

Twice through my bedroom window
I've seen the horned owl drop from the oaks to panic
the rabbit in my neighbor's backyard.
Last night he paced for an hour across the top
of the cage, scrutinizing
the can of water, the mound of pellets,
turning his genius to the riddle
of the wire, while under him
the rabbit balled like a fat carnation in the wind.

Both of the terriers yapped from their porch
but the owl never flinched, pacing,
clawing the wire, spreading wings like a gray cape,
leaping, straining to lift the whole cage,
and the cage rocking
on its stilts, settling, and rocking again,
until he settled with it, paused,
and returned to a thought.

And the rabbit, ignorant of mercy,
curled on itself in that white drift
of feathers?

Wait, three years and I haven't escaped the child
I saw at Northside the night
my daughter was born,

a little brown sack of twigs
curled under glass, eyes bulging,
trembling in the monitors,
and the nurses
rolling the newborns out to nurse,
and the shadows sweeping the nursery.

DAVID BOTTOMS

Great Gray Owl

The Great Gray Owl is the tallest American owl, two to three feet high, with the widest wingspan, from more than four up to five feet, but it has small feet for its size and preys on only small rodents. It is native to the boreal forests of North America and Eurasia.

Great Gray Owl

Who knew you would grow from gray bark
So that nothing is separate or new
But your yellow eyes following through
From the mottling brown in the dark,
Spectral Owl—from the spiral, the spark
That the circling feathers lead to?
Who knew you could speak as you do,
Great Gray Ghost—who knew you could speak?

ANNIE FINCH

The Owl

His element is silent and inexorable.
Mack the Knife waits in his eyes,
yet he is generous and brings his young
eleven mice four bullheads
thirteen grouse two eels
three rabbits and a woodcock
all in one night.

Is it too much to expect prose
to learn from the owl
his exact knowledge of his object,
his exact eyes claws wings
and be the scourge of rats?
It might, like him, then live to
sixty-eight years in the clear impersonal
and look wise and imperturbable.

CARL RAKOSI

The Owl

Downhill I came, hungry, and yet not starved,
Cold, yet had heat within me that was proof
Against the north wind; tired, yet so that rest
Had seemed the sweetest thing under a roof.

Then at the inn I had food, fire, and rest,
Knowing how hungry, cold, and tired was I.
All of the night was quite barred out except
An owl's cry, a most melancholy cry.

Shaken out long and clear upon the hill
No merry note, nor cause of merriment,
But one telling me plain what I escaped
And others could not, that night, as in I went.

And salted was my food, and my repose,
Salted and sobered too, by the bird's voice
Speaking for all who lay under the stars,
Soldiers and poor, unable to rejoice.

DAS 2005

Chimney Swift

The Chimney Swift is almost constantly in flight, catching and eating small insects on the wing and stopping only to nest or to roost at night. Frequent banking and turning can give the illusion that swifts beat one wing at a time. Swifts are found worldwide, although the range of the Chimney Swift is from eastern North America to western South America.

Throughout the winter, we once believed, they hid
    Nearby us, under eaves,
In nestlike thatch and thickets wedged in tile,
Sleeping as close to us as figures carved
    On vaults and open rafters.

They were, in fact, skimming the Amazon

They are back now, with cowbirds, boat-tailed grackles,
    Kingbirds on powerlines,
And quick goldfinches heading for the fields
They drown their color in, in northern mountains.

    Swifts funnel down at twilight
Into cold flues, chattering like children.

They speak their language and we listen
    In our own, comparing them
To children, travellers, speed, and life itself,
Imparting a charmed knowledge unto us.

    To them, there are two worlds—
The soot-thick shaft and the silky bowl of sky.

To watch for them, to become expectant,
    To need their spring arrival,
To know the kink from craning back the neck
During the warm, late afternoons of April,
    Is part of the enchantment,

Is to believe they feel it, too, and act.

MARK JARMAN

Spring comes little, a little. All April it rains.
The new leaves stick in their fists; new ferns still fiddleheads.
But one day the swifts are back. Face to the sun like a child
You shout, "The swifts are back!"

Sure enough, bolt nocks bow to carry one sky-scyther
Two hundred miles an hour across fullblown windfields.
Swereee swereee. Another. And another.
It's the cut air falling in shrieks on our chimneys and roofs.

The next day, a fleet of high crosses cruises in ether.
These are the air pilgrims, pilots of air rivers.
But a shift of wing, and they're earth-skimmers, daggers
Skilful in guiding the throw of themselves away from themselves.

Quick flutter, a scimitar upsweep, out of danger of touch, for
Earth is forbidden to them, water's forbidden to them,
All air and fire, little owlish ascetics, they outfly storms,
They rush to the pillars of altitude, the thermal fountains.

Here is a legend of swifts, a parable
When the Great Raven bent over earth to create the birds,
The swifts were ungrateful. They were small muddy things
Like shoes, with long legs and short wings,

So they took themselves off to the mountains to sulk.
And they stayed there. "Well," said the Raven, after years of this,
"I will give you the sky. You can have the whole sky
On condition that you give up the rest."

"Yes, yes," screamed the swifts, "We abhor the rest.
We detest the filth of growth, the sweat of sleep,
Soft nests in the wet fields, slimehold of worms.
Let us be free, be air!"

So the Raven took their legs and bound them into their bodies.
He bent their wings like boomerangs, honed them like knives.
He streamlined their feathers and stripped them of velvet.
Then he released them, *Never to Return*

Inscribed on their feet and wings. And so
We have swifts, though in reality, not parables but
Bolts in the world's need: swift
Swifts, not in punishment, not in ecstasy, simply
Sleepers over oceans in the mill of the world's breathing.
The grace to say they live in another firmament.
A way to say the miracle will not occur,
And watch the miracle.

ANNE STEVENSON

Swifts

Bing Crosby died in Spain
while playing golf with Franco
but who could care less, and at this
writing only a few of
my dear ones are gone—ah I
could make a sad list—the swifts,
as if to prove a point,
fly into the light and make
a mockery out of our darkness.
They scream for food but in
the world of shadows they only
make a quick motion; I have
studied them—the whiter
the wall is—the barer the bulb—
the more they scream, the more
they dip down. I have made
my two hands into a shape
and I have darkened the wall
to see what it looks like—I have
shortened my two broken fingers
to make the small tall and twisted
the knuckles sideways so when
they come in to eat one shadow
overtakes the other, that way

I can live in the darkness
with Franco's poisonous head
and Crosby's ears, who fainted,
a thousand to one, behind a
number two club, though no swift
died for him, well, for them,
digging for clubs. I watch the
birds every night; they fly
In a great circle, much larger
than what I can see, their dipping
Is what I dreaded in front of
my plain white wall—I say it
for the nine hundred Americans
who died in Spain. I thought
I'd have to wait forever
to do them a tiny justice
and listen to their songs
and die a little from the foolhardy
mournful words, flying down
one air current or another
and doing the sides of buildings
and tops of trees, the low-lying
straggling dogwood, the full-bodied
huge red maple, my dear ones.

Ruby-throated Hummingbird

The Ruby-throated Hummingbird is the only species of hummingbird that breeds in the eastern half of North America. Powerful in spite of its small size, and with a wing beat of more than fifty times a second, it flies without stopping over the Gulf of Mexico to winter in southern Mexico and Central America.

Humming-Bird

I can imagine, in some otherworld
Primeval-dumb, far back
In that most awful stillness, that only gasped and hummed,
Humming-birds raced down the avenues.

Before anything had a soul,
While life was a heave of Matter, half inanimate,
This little bit chipped off in brilliance
And went whizzing through the slow, vast, succulent stems.

I believe there were no flowers then,
In the world where the humming-bird flashed ahead of creation.
I believe he pierced the slow vegetable veins with his long beak.

Probably he was big
As mosses, and little lizards, they say, were once big.
Probably he was a jabbing, terrifying monster.
We look at him through the wrong end of the long telescope of Time,
Luckily for us.

D. H. LAWRENCE

DAS 2005

Belted Kingfisher

Unusually for birds, the female Belted Kingfisher is more brightly colored than the male. Common throughout North America, it nests in burrows, usually in a dirt bank near water. A pair of kingfishers will defend their territory (about half a mile along a stream) from other kingfishers.

The Kingfisher

I wanted to see a kingfisher
with its throat bound up in whiteness
and its black crest aimed at clouds.

I didn't know what it looked like,
not really. In poems and stories
it would flicker, a subtle omen.

But a kingfisher appeared
one February Sunday.
First, a high, rattling call

like a constant shake of maracas.
Then the bird itself touched down
on an aged tree, on a pond's island,

in a circle of melting ice.
From that one place, it called and called
and its call tapped a contradiction

to the cold, a noise that loosened
the ice's thin sheets.
The kingfisher lifted its tail

up and down, moved close to the water,
moved closer. Its eyes skimmed the pond.
I clumsily focused binoculars:

the white throat, the angular crest!
—perceptible, barely, by color
and form, a lot like a painting

viewed so close up it's blurred.
Step away. Step away. I didn't
from my life's one mention of kingfisher

until some noise
(a rifle, or muffler, or tree fall
in the distance) triggered its flight

and then I watched it lift
—it's heavy, a bird more burdened
than some, and not all grace—

trailing calls like the beads of a rosary:
a string of clicks in air,
a shadow leaving the ice.

The Kingfisher

In a year the nightingales were said to be so loud
they drowned out slumber, and peafowl strolled screaming
beside the ruined nunnery, through the long evening
of a dazzled pub crawl, the halcyon color, portholed
by those eye-spots' stunning tapestry, unsettled
the pastoral nightfall with amazements opening.

Months later, intermission in a pub on Fifty-fifth Street
found one of them still breathless, the other quizzical,
acting the philistine, puncturing Stravinsky—"Tell
me, what *was* that racket in the orchestra about?"—
hauling down the Firebird, harum-scarum, like a kite,
a burnished, breathing wreck that didn't hurt at all.

Among the Bronx Zoo's exiled jungle fowl, they heard
through headphones of a separating panic, the bellbird
reiterate its single *chong*, a scream nobody answered.
When he mourned, "The poetry is gone," she quailed,
seeing how his hands shook, sobered into feeling old.
By midnight, yet another fifth would have been killed.

A Sunday morning, the November of their cataclysm
(Dylan Thomas brought in *in extremis* to St. Vincent's,
that same week, a symptomatic datum) found them
wandering a downtown churchyard. Among its headstones,
while from unruined choirs the noise of Christendom
poured over Wall Street, a benison in vestments,

a late thrush paused, in transit from some grizzled
spruce bog to the humid equatorial fireside: berry-
eyed, bark-brown above, with dark hints of trauma
in the stigmata of its underparts—or so, too bruised
just then to have invented anything so fancy,
later, re-embroidering a retrospect, she had supposed.

In gray England, years of muted recrimination (then
dead silence) later, she could not have said how many
spoiled takeoffs, how many entanglements gone sodden,
how many gaudy evenings made frantic by just one
insomniac nightingale, how many liaisons gone down
screaming in a stroll beside the ruined nunnery;

a kingfisher's burnished plunge, the color
of felicity afire, came glancing like an arrow
through landscapes of untended memory: ardor
illuminating with its terrifying currency
now no mere glimpse, no porthole vista
but, down on down, the uninhabitable sorrow.

AMY CLAMPITT

Pileated Woodpecker

The Pileated Woodpecker chips out rectangular holes in trees that are so large that other birds feed in the excavations and small trees may split. A pair of these large woodpeckers stay together on their territory year round. It is found across Canada and along parts of the Pacific Coast and the eastern United States.

The Woodpecker Keeps Returning

The woodpecker keeps returning
to drill the house wall.
Put a pie plate over one place, he chooses another.

There is nothing good to eat there:
he has found in the house
a resonant billboard to post his intentions,
his voluble strength as provider.

But where is the female he drums for? Where?

I ask this, who am myself the ruined siding,
the handsome red-capped bird, the missing mate.

The Woodpecker

The woodpecker pecked out a little round hole
And made him a house in the telephone pole.
One day when I watched he poked out his head,
And he had on a hood and a collar of red.

When the streams of rain pour out of the sky,
And the sparkles of lightning go flashing by,
And the big, big wheels of thunder roll,
He can snuggle back in the telephone pole.

ELIZABETH MADOX ROBERTS

Scissor-tailed Flycatcher

The Great Crested Flycatcher in the poem is a more modest-looking bird than this spectacular Scissor-tailed Flycatcher. Natives of North America, the Great Crested is an eastern flycatcher; the Scissor-tailed, midwestern. Both like to line their nests with interesting objects: the Great Crested with snakeskin, and the Scissor-tailed with human debris such as string, cloth, carpet lint, and cigarette butts.

Now I have the Great Crested Flycatcher
amidst my Red Delicious, the tree's
spindly arm so freighted with apples
it sags under the bird's bird-weight
then springs at his departure
like the board just after a diver's flung up
and gone. *Weep weep weep*, he trills
from the overgrown fence row,
his three notes so laden with gravity
I wonder is this song or his lament,
one wing among the green going going?

And that, my friends, is how reason
insinuates its bone lonely self
among the arts of joy—the least of which
is knowing when to snip the string
that tethers us, our sky blue why.

The bird's after-image is more than
I can take, really, more than I can ask
of Wednesday's usual desultory coffers,
high noon offering its unspent zenith.
I want to say there's absolutely nothing
like this vision of bird and apples. I want
to say absolutely nothing else gives
of wings and fruit. Then I think of
nights my wife rose flushed above me—
this, the only store I put in absolutes.

· · · · · · · · · · · · · · · · · · · ·

KEVIN STEIN

Eastern Phoebe

Phoebes are in the habit of dipping or wagging their tails, like the bird in the poem. Their call sounds like their name: *phee-beee*. This Eastern Phoebe differs from its western counterpart in the poem, the Black Phoebe, in that it does migrate to the southern United States and Mexico for the winter.

Black Phoebe

Her swoops are short and low and don't aspire
To more, it seems, than nature's common strife.
Perching, she strops her bill upon a wire
As though she'd barbered in a former life.
When the wire rocks, she quickly dips her tail
A few times, and her balance doesn't fail.

If she displays an unassuming pride—
Compact, black-capped, black breast puffed to the sun—
The sentiment perhaps is justified:
Mosquitoes, gnats, and flies would overrun
Much of the planet within several years
But for her and her insectivorous peers.

Not prone, as are the jays, to talking trash,
She offers quieter companionship;
On summer days, when starlings flap and splash
And make the birdbath overspill and drip
Or empty out its basin altogether,
She seeks the shade and waits for cooler weather.

When autumn whips the plum tree to and fro
And rains slick its smooth trunk, and pools collect
Among its exposed roots, and Mexico
Tempts most birds of the garden to defect,
It is a cheering check against chagrin
To think this is the place she'll winter in.

She makes, for now, a series of abrupt
Dives, lifts, and turns, and lights upon a stake.
A moth flaps past; she darts to interrupt
Its course and then retrieves her perch to make
A thorough survey, though at no great height,
Of plants confided to her oversight.

TIMOTHY STEELE

Northern Shrike

The black-masked Northern Shrike, a songbird, feeds on small birds, small mammals, and insects. It has been observed to kill more than it can eat, storing excess food for times of scarcity.

The Shrike

Hark—hark—from out the thickest fog
Warbles with might and main
The fearless shrike, as all agog
To find in fog his gain.

His steady sail he never furls
At any time o' year,
And perched now on winter's curls,
He whistles in his ear.

Blue Jay

Among its many calls (one of them described as a jeer), the Blue Jay mimics those of hawks, especially the Red-shouldered Hawk, perhaps to warn that the raptor is in the vicinity or perhaps to fool the other birds—leaving more insects and nuts for the taking.

Still Missing the Jays

Then this afternoon, in the anonymous
winter hedge, I saw one. I'd just climbed,
in my sixty-year-old body—with its heart
attacks, kidney stones, torn Achilles tendon,
vague promises of ulcers, various subtle,
several visible permanent scars, ghost-
gray hair, long nights and longer silences,
impotence and liver spots, evident
translucence, sometime short-term memory loss—
I'd just climbed out of the car and there
it was, eye-level, looking at me, young,
bare blue, the crest and marking jewelry
penciled in, smaller than it would be
if it lasted but large enough to show
the dark adult and make its queedle
and complaint. It seemed to wait for me,
watching in that superciliary way
birds watch too. So I took it as a sign,
part spring, part survival. I hadn't seen a jay
in years—I'd almost forgotten they existed.
Such obvious, quarrelsome, vivid birds
that turn the air around them crystalline.
Such crows, such ravens, such magpies!
Such bristling in the spyglass of the sun.
Yet this one, new in the world,
softer, plainer, curious. I tried
to match its patience, not to move,
though when it disappeared to higher ground,
I had the thought that if I opened up my hand—

STANLEY PLUMLY

Black-billed Magpie

The Black-billed Magpie, common in western North America, was thought to
be the same species as the Eurasian Magpie. Studies of the vocal and behavioral
differences between the birds, however, have led ornithologists to believe that
the Black-billed Magpie is more closely related to the Yellow-billed Magpie, found
only in southern California, than to the black-billed Eurasian Magpie, native to
Europe, North Africa, and much of Asia.

Magpie's Song

Six A.M.,
Sat down on excavation gravel
by juniper and desert S.P. tracks
interstate 80 not far off
between trucks
Coyotes—maybe three
howling and yapping from a rise.

Magpie on a bough
Tipped his head and said,

"Here in the mind, brother
Turquoise blue.
I wouldn't fool you.
Smell the breeze
It came through all the trees
No need to fear
What's ahead
Snow up on the hills west
Will be there every year
be at rest.
A feather on the ground—
The wind sound—

Here in the Mind, Brother,
Turquoise Blue"

GARY SNYDER

American Crow

Most young American Crows do not breed until they are at least four years old, and many stay with their parents and help them raise their younger siblings. A family of crows can number up to fifteen, with young from five years, and wintertime communal roosts can include up to hundreds of thousands of birds.

Paired Things

Who, who had only seen wings,
could extrapolate the
skinny sticks of things
birds use for land,
the backward way they bend,
the silly way they stand?
And who, only studying
birdtracks in the sand,
could think those little forks
had decamped on the wind?
So many paired things seem odd.
Who ever would have dreamed
the broad winged raven of despair
would quit the air and go
bandylegged upon the ground,
a common crow?

KAY RYAN

The Calves Not Chosen

The mind goes *caw, caw, caw, caw,*
dark and fast. The orphan heart
cries out, "Save me. Purchase me
as the sun makes the fruit ripe.
I am one with them and cannot feed
on winter dawns." The black birds
are wrangling in the fields
and have no kindness, all sinew
and stick bones. Both male and female.
Their eyes are careless of cold and rain,
of both day and night. They love nothing
and are murderous with each other.
All things of the world are bowing
or being taken away. Only a few calves
will be chosen, the rest sold for meat.
The sound of the wind grows bigger
than the tree it's in, lessens only
to increase. *Haw, haw,* the crows call,
awake or asleep, in white, in black.

LINDA GREGG

The Crows Start Demanding Royalties

Of all the birds, they are the ones
who mind their being armless most:
witness how, when they walk, their heads jerk
back and forth like rifle bolts.
How they heave their shoulders into each stride
as if they hoped that by some chance
new bones there would come popping out
with a boxing glove on the end of each.

Little Elvises, the hairdo slicked
with too much grease, they convene on my lawn
to strategize for their class-action suit.
Flight they would trade in a New York minute
for a black muscle car and a fist on the shift
at any stale green light. But here in my yard
by the Jack-in-the-Box Dumpster
they can only fossick in the grass for remnants

of the world's stale buns. And this
despite all the crow poems that have been written
because men like to see themselves as crows
(the head-jerk performed in the rearview mirror,
the dark brow commanding the rainy weather).
So I think I know how they must feel:
ripped off, shook down, taken to the cleaners.
What they'd like to do now is smash a phone against a wall.
But they can't, so each one flies to a bare branch and screams.

LUCIA PERILLO

The Questions Poems Ask

Watching a couple of crows
playing around in the woods, swooping
in low after each other, I wonder
if they ever slam into the trees.

There's an answer here, unlike
most questions in poems,
which are left up in the air.
Was it a vision or a waking dream?

You decide, says the poet.
You do some of this work,
but think carefully.
Some people want to believe

poetry is anything
they happen to feel. That way
they're never wrong. Others yearn
for the difficult:

insoluble problems, secret codes
not meant to be broken.
Nobody, they've discovered,
ever means what he says.

But rarely does a crow
hit a tree, though other, clumsier birds
bang into them all the time, and we say
these birds have not adapted well

to the forest environment.
Frequently stunned, they become
easy prey for the wily fox,
who's learned how to listen

for that snapping of branches
and collapsing of wings,
who knows where to go
and what to do when he gets there.

· · · · · · · · · · · · · · · · · · ·

Common Raven

Crows and ravens are known as songbirds, despite their hoarse calls. The Common Raven is the largest of the songbirds. It is found throughout the Northern Hemisphere—from Arctic to desert environments—and is perhaps the smartest of all birds.

Ravens at Deer Creek

Something's dead in that stand of fir
one ridge over. Ravens circle and swoop
above the trees, while others
swirl up from below, like paper scraps
blackened in a fire. In the mountains
in winter, it's true: death is a joyful flame,
those caws and cartwheels pure celebration.
It is a long snowy mile I've come
to see this, thanks to dumb luck or grace.
I meant only a hard ski through powder,
my pulse in my ears, and sweat, the pace
like a mainspring, my breath louder and louder
until I stopped, body an engine
ticking to be cool. And now the birds.
I watch them and think, maybe I have seen
these very ones, speaking without words,
cleared-eyed and clerical, ironic, peering in at me
from the berm of snow outside my window,
where I sprinkled a few crumbs of bread. We
are neighbors in the neighborhood of silence.

They've accepted my crumbs, and when the fire was hot
and smokeless huddled in ranks against
the cold at the top of the chimney. And they're not
without gratitude. Though I'm clearly visible
to them now, they swirl on and sing,
and if, in the early dusk, I should fall
on my way back home and—injured, weeping—
rail against the stars and frigid night
and crawl a while on my hopeless way
then stop, numb, easing into the darkening white
like a candle, I know they'll stay
with me, keeping watch, moving limb to limb,
angels down Jacob's ladder, wise
to the moon, and waiting for me, simple as sin,
that they may know the delicacy of my eyes.

ROBERT WRIGLEY

The Ravens of Denali

Such dumb luck. To stumble
across an "unkindness" of ravens
at play with a shred of clear visquine
fallen from the blown-out window
of the Denali Truck Stop and Café.
Black wings gathering in the deserted
parking lot below the Assembly of God.
Ravens at play in the desolate fields
of the lord, under the tallest mountain
in North America, eight of them,
as many as the stars in the Big Dipper
on Alaska's state flag, yellow stars
sewn to a blue background flapping
from a pole over the roadside.
Flag that Benny Benson, age 13,
an Alutiiq Indian of Seward
formerly housed at the Jesse-Lee Memorial
Home for Orphans in Unalaska,
designed and submitted to a contest
in 1927 and won, his crayoned masterpiece
snapping above every broken-down
courthouse, chipped brick library
and deathtrap post office
in the penultimate state accepted
to the Union, known to its people
as the Upper One. Though a design

of the northern lights would have been
my choice, those alien green curtains
swirling over Mt. McKinley, Denali,
"the tall one," during the coldest, darkest
months of the subarctic year.
Red starburst or purple-edged skirt
rolling in vitreous waves
over the stunted ice-rimed treetops
or in spring, candles of fireweed
and the tiny ice blue flowers
of the tundra. Tundra, a word
that sounds like a thousand caribou
pouring down a gorge.
But all that might be difficult
for an orphaned 7th grader to draw
with three chewed-up crayons
and a piece of butcher paper.
As would these eight giggling ravens
with their shrewd eyes and silt-shine wings,
beaks like keloid scars. Acrobats
of speed and sheen. Black boot
of the bird family. Unconcerned
this moment with survival.
Though I hope they survive.
Whatever we have in store for them.
And the grizzly bear and the club-

footed moose. The muscular salmon.
The oil-spill seal and gull.
And raven's cousin, the bald eagle,
who can dive at 100 miles per hour,
can actually swim with massive
butterfly strokes through
the great glacial lakes of Alaska,
her wingspan as long as a man.
Architect of the two-ton nest
assembled over 34 years
with scavenged branches,
threatened in all but three
of the Lower 48, but making, by god,
a comeback if it's not too late
for such lofty promises.
Even the homely marmot
and the immigrant starling,
I wish you luck,
whatever ultimate harm we do
to this northernmost up-flung arm
of our country, our revolving world.
But you, epicurean raven, may you
be the pole star of the apocalypse,
you stubborn snow trudger,
you quorum of eight who jostle one another
for a strip of plastic on the last

· · · · · · · · · · · · · · · · · · · ·

endless day, the last endless night
of our only sun's solar wind,
those glorious auroras, glassine gowns
of Blake's angels, that almost invisible shine
tugged and stretched between you
like taffy from outer space, tattered ends
gripped in your fur-crusted beaks as we reel
headlong into the dwindling unknown.
Denizens of the frozen north, the last
frontier, harbingers of unluck,
and the cold bleak lack to come.

DORIANNE LAUX

Barn Swallow

The Barn Swallow is one of the most abundant and wide-ranging swallow species in the world, breeding throughout the Northern Hemisphere and wintering in much of the Southern Hemisphere. It is easily identified by its long, forked tail.

Ode to the Swallow

Thou indeed, little Swallow,
A sweet yearly comer,
Art building a hollow
New nest every summer,
And straight dost depart
Where no gazing can follow,
Past Memphis, down Nile!
Ah! but Love all the while
Builds his nest in my heart,
Through the cold winter-weeks:
And as one Love takes flight,
Comes another, O Swallow,
In an egg warm and white,
And another is callow.
And the large gaping beaks
Chirp all day and all night:
And the Loves who are older
Help the young and the poor Loves,
And the young Loves grown bolder
Increase by the score Loves—
Why, what can be done?
If a noise comes from one
Can I bear all this rout of a hundred and more Loves?

The Swallow

1

Pretty Swallow once again
Come and pass me i' the rain
Pretty swallow why so shy
Pass again my window by.

2

The horse pond where he dips his wings
The wet day prints it full o' rings
The raindrops on his track
Lodge like pearls upon his back.

3

Then again he dips his wing
In the wrinkles of the spring
Then o'er the rushes flies again
And pearls roll off his back like rain.

4

Pretty little swallows fly
Village doors and windows by
Whisking o'er the garden pales
Where the blackbird finds the snails.

5

Whewing by the ladslove tree
For something only seen by thee
Pearls that on the red rose hings
Falls off shaken by thy wings.

6

On yon low thatched cottage stop
In the sooty chimney pop
Where thy wife and family
Every evening wait for thee.

The Emperor's Bird's-Nest

Once the Emperor Charles of Spain,
    With his swarthy, grave commanders,
I forget in what campaign,
Long besieged, in mud and rain,
    Some old frontier town of Flanders.

Up and down the dreary camp,
    In great boots of Spanish leather,
Striding with a measured tramp,
These Hidalgos, dull and damp,
    Cursed the Frenchmen, cursed the weather.

Thus as to and fro they went,
    Over upland and through hollow,
Giving their impatience vent,
Perched upon the Emperor's tent,
    In her nest, they spied a swallow.

Yes, it was a swallow's nest,
    Built of clay and hair of horses,
Mane, or tail, or dragoon's crest,
Found on hedge-rows east and west,
    After skirmish of the forces.

Then an old Hidalgo said,
    As he twirled his gray mustachio,
"Sure this swallow overhead
Thinks the Emperor's tent a shed,
    And the Emperor but a Macho!"

Hearing his imperial name
    Coupled with those words of malice,
Half in anger, half in shame,
Forth the great campaigner came
    Slowly from his canvas palace.

"Let no hand the bird molest,"
    Said he solemnly, "nor hurt her!"
Adding then, by way of jest,
"Golondrina is my guest,
    'Tis the wife of some deserter!"

Swift as bowstring speeds a shaft,
    Through the camp was spread the rumor,
And the soldiers, as they quaffed
Flemish beer at dinner, laughed
    At the Emperor's pleasant humor.

. . . .

So unharmed and unafraid
    Sat the swallow still and brooded,
Till the constant cannonade
Through the walls a breach had made,
    And the siege was thus concluded.

Then the army, elsewhere bent,
    Struck its tents as if disbanding,
Only not the Emperor's tent,
For he ordered, ere he went,
    Very curtly, "Leave it standing!"

So it stood there all alone,
    Loosely flapping, torn and tattered,
Till the brood was fledged and flown,
Singing o'er those walls of stone
    Which the cannon-shot had shattered.

Watching the swallows

Watching the swallows
That flew about restlessly,
And flung their shadows
Upon the sunbright walls of the old building;
The shadows glanced and twinkled,
Interchanged and crossed each other,
Expanded and shrunk up,
Appeared and disappeared, every instant;
As I observed to William and Coleridge,
Seeming more like living things
Than the birds themselves.

DOROTHY WORDSWORTH

from The Princess; O Swallow

O Swallow, Swallow, flying, flying South,
Fly to her, and fall upon her gilded eaves,
And tell her, tell her, what I tell to thee.

O tell her, Swallow, thou that knowest each,
That bright and fierce and fickle is the South,
And dark and true and tender is the North.

O Swallow, Swallow, if I could follow, and light
Upon her lattice, I would pipe and trill,
And cheep and twitter twenty million loves.

O were I thou that she might take me in,
And lay me on her bosom, and her heart
Would rock the snowy cradle till I died.

Why lingereth she to clothe her heart with love,
Delaying as the tender ash delays
To clothe herself, when all the woods are green?

O tell her, Swallow, that thy brood is flown:
Say to her, I do but wanton in the South,
But in the North long since my nest is made.

O tell her, brief is life but love is long,
And brief the sun of summer in the North,
And brief the moon of beauty in the South.

O Swallow, flying from the golden woods,
Fly to her, and pipe and woo her, and make her mine,
And tell her, tell her, that I follow thee.

. . . .

ALFRED, LORD TENNYSON

Home-Thoughts, from Abroad

I

Oh, to be in England
Now that April's there,
And whoever wakes in England
Sees, some morning, unaware,
That the lowest boughs and the brushwood sheaf
Round the elm-tree bole are in tiny leaf,
While the chaffinch sings on the orchard bough
In England—now!

II

And after April, when May follows,
And the whitethroat builds, and all the swallows!
Hark, where my blossomed pear-tree in the hedge
Leans to the field and scatters on the clover
Blossoms and dewdrops—at the bent spray's edge—
That's the wise thrush; he sings each song twice over,
Lest you should think he never could recapture
The first fine careless rapture!
And though the fields look rough with hoary dew,
All will be gay when noontide wakes anew
The buttercups, the little children's dower
—Far brighter than this gaudy melon-flower!

ROBERT BROWNING

Cliff Swallow

The Cliff Swallow breeds in North America and winters in South America. Its gourd-shaped nest, built on the vertical wall of a cliff or a building, usually under an overhang, is made of mud pellets and is lined with grass.

Is it some turn of wind
that funnels them all down at once, or
is it their own voices netting
to bring them in—the roll and churr
of hundreds searing through river light
and cliff dust, each to its precise
mud nest on the face
none of our own isolate
groping, wishing need could be sent
so unerringly to solace. But
this silk-skein flashing is like heaven
brought down: not to meet ground
or water—to enter
the riven earth and disappear.

DEBRA NYSTROM

The Blue Swallows

Across the millstream below the bridge
Seven blue swallows divide the air
In shapes invisible and evanescent,
Kaleidoscopic beyond the mind's
Or memory's power to keep them there.

"History is where tensions were,"
"Form is the diagram of forces."
Thus, helplessly, there on the bridge,
While gazing down upon those birds—
How strange, to be above the birds!—
Thus helplessly the mind in its brain
Weaves up relation's spindrift web,
Seeing the swallows' tails as nibs
Dipped in invisible ink, writing . . .

Poor mind, what would you have them write?
Some cabalistic history
Whose authorship you might ascribe
To God? to Nature? Ah, poor ghost,
You've capitalized your Self enough.
That villainous William of Occam
Cut out the feet from under that dream
Some seven centuries ago.
It's taken that long for the mind
To waken, yawn and stretch, to see
With opened eyes emptied of speech
The real world where the spelling mind
Imposes with its grammar book
Unreal relations on the blue
Swallows. Perhaps when you will have
Fully awakened, I shall show you
A new thing: even the water
Flowing away beneath those birds
Will fail to reflect their flying forms,
And the eyes that see become as stones
Whence never tears shall fall again.

O swallows, swallows, poems are not
The point. Finding again the world,
That is the point, where loveliness
Adorns intelligible things
Because the mind's eye lit the sun.

. . . .

HOWARD NEMEROV

DAS 2006

Black-capped Chickadee

The calls of this beloved bird of northern North America—the most familiar sounding out its name, *chick-a-dee-dee-dee*—have been found to be complex and language-like, codes giving predator alarms and other contact calls. A familiar visitor to bird feeders, the Black-capped Chickadee may store seeds in different locations, remembering thousands of caches when it comes time to recover the seeds.

December Notes

The backyard is one white sheet
Where we read in the bird tracks

The songs we hear. Delicate
Sparrow, heavier cardinal,

Filigree threads of chickadee.
And wing patterns where one flew

Low, then up and away, gone
To the woods but calling out

Clearly its bright epigrams.
More snow promised for tonight.

The postal van is stalled
In the road again, the mail

Will be late and any good news
Will reach us by hand.

NANCY MCCLEERY

In Walden wood the chickadee

In Walden wood the chickadee
Runs round the pine and maple tree,
Intent on insect slaughter:
O tufted entomologist!

Devour as many as you list,
Then drink in Walden Water.

RALPH WALDO EMERSON

DAS 2003

Tufted Titmouse

The name titmouse dates to the fourteenth century, from *māse*, the Old English word for "bird," and *tit*, which referred to something small. In Great Britain, Great Tits and Blue Tits were reported to have broken open the foil caps of milk bottles, delivered to doorsteps, to drink the cream. The Tufted Titmouse is a year-round resident of the eastern half of the United States, often seen at bird feeders and in wooded areas.

If you would happy company win,
Dangle a palm-nut from a tree,
Idly in green to sway and spin,
Its snow-pulped kernel for bait; and see
    A nimble titmouse enter in.

Out of earth's vast unknown of air,
Out of all summer, from wave to wave,
He'll perch, and prank his feathers fair,
Jangle a glass-clear wildering stave,
    And take his commons there—

This tiny son of life; this spright,
By momentary Human sought,
Plume will his wing in the dappling light,
Clash timbrel shrill and gay—
And into time's enormous nought,
    Sweet-fed, will flit away.

Eastern Bluebird

The Eastern and the Western Bluebird are closely related. Medium-size thrushes found in open woodlands, farmlands, and suburban birdhouses, they are native to North America. The population decline of the Eastern Bluebird in the mid-twentieth century—the result of habitat loss, pesticide poisoning, and nest predation by the European Starling and the House Sparrow—was reversed as a result of the building and monitoring of nest boxes.

The Bluebirds

In the midst of the poplar that stands by our door,
We planted a bluebird box,
And we hoped before the summer was o'er
A transient pair to coax.

One warm summer's day the bluebirds came
And lighted on our tree,
But at first the wand'rers were not so tame
But they were afraid of me.

They seemed to come from the distant south,
Just over the Walden wood,
And they skimmed it along with open mouth
Close by where the bellows stood.

Warbling they swept round the distant cliff,
And they warbled it over the lea,
And over the blacksmith's shop in a jiff
Did they come warbling to me.

They came and sat on the box's top
Without looking into the hole,
And only from this side to that did they hop,
As 'twere a common well-pole.

Methinks I had never seen them before,
Nor indeed had they seen me,
Till I chanced to stand by our back door,
And they came to the poplar tree.

In course of time they built their nest
And reared a happy brood,
And every morn they piped their best
As they flew away to the wood.

Thus wore the summer hours away
To the bluebirds and to me,
And every hour was a summer's day,
So pleasantly lived we.

They were a world within themselves,
And I a world in me,
Up in the tree—the little elves—
With their callow family.

One morn the wind blowed cold and strong,
And the leaves when whirling away;
The birds prepared for their journey long
That raw and gusty day.

Boreas came blust'ring down from the north,
And ruffled their azure smocks,
So they launched them forth, though somewhat loth,
By way of the old Cliff rocks.

Meanwhile the earth jogged steadily on
In her mantle of purest white,
And anon another spring was born
When winter was vanished quite.

And I wandered forth o'er the steamy earth,
And gazed at the mellow sky,
But never before from the hour of my birth
Had I wandered so thoughtfully.

For never before was the earth so still,
And never so mild was the sky,
The river, the fields, the woods, and the hill,
Seemed to heave an audible sigh.

I felt that the heavens were all around,
And the earth was all below,
As when in the ears there rushes a sound
Which thrills you from top to toe.

. . . .

I dreamed that I was a waking thought—
A something I hardly knew—
Not a solid piece, nor an empty nought,
But a drop of morning dew.

'Twas the world and I at a game of bo-peep,
As a man would dodge his shadow,
An idea becalmed in eternity's deep—
'Tween Lima and Segraddo.

Anon a faintly warbled note
From out the azure deep,
Into my ears did gently float
As is the approach of sleep.

It thrilled but startled not my soul;
Across my mind strange mem'ries gleamed,
As often distant scenes unroll
When we have lately dreamed.

The bluebird had come from the distant South
To his box in the poplar tree,
And he opened wide his slender mouth,
On purpose to sing to me.

HENRY DAVID THOREAU

The Darkling Thrush

I leant upon a coppice gate,
 When Frost was spectre-gray,
And Winter's dregs made desolate
 The weakening eye of day.
The tangled bine-stems scored the sky
 Like strings of broken lyres,
And all mankind that haunted nigh
 Had sought their household fires.

The land's sharp features seemed to be
 The Century's corpse outleant,
Its crypt the cloudy canopy,
 The wind its death-lament.
The ancient pulse of germ and birth
 Was shrunken hard and dry,
And every spirit upon earth
 Seemed fervorless as I.

At once a voice arose among
 The bleak twigs overhead,
In a full-hearted evensong
 Of joy illimited.
An aged thrush, frail, gaunt and small,
 In blast-beruffled plume,
Had chosen thus to fling his soul
 Upon the growing gloom.

So little cause for carolings
     Of such ecstatic sound
Was written on terrestrial things
     Afar or nigh around,
That I could think there trembled through
     His happy good-night air
Some blessed Hope, whereof he knew
     And I was unaware.

THOMAS HARDY

Tossed on the glittering air they soar and skim,
Whose voices make the emptiness of light
A windy palace. Quavering from the brim
Of dawn, and bold with song at edge of night,
They clutch their leafy pinnacles and sing
Scornful of man, and from his toils aloof
Whose heart's a haunted woodland whispering;
Whose thoughts return on tempest-baffled wing;
Who hears the cry of God in everything,
And storms the gate of nothingness for proof.

SIEGFRIED SASSOON

American Robin

The American Robin is the most widespread thrush in North America. It is not related to the much smaller European Robin, which is now considered to be an Old World flycatcher. Although it is associated with the arrival of spring, the robin is a year-round resident of much of its breeding range but spends the winter in flocks, not in backyards.

I have a Bird in spring

I have a Bird in spring
Which for myself doth sing—
The spring decoys.
And as the summer nears—
And as the rose appears,
Robin is gone.

Yet I do not repine
Knowing that Bird of mine
Though flown—
Learneth beyond the sea
Melody new for me
And will return.

Fast in a safer hand—
Held in a truer Land
Are mine—
And though they now depart,
Tell I my doubting heart
They're thine.

In a serener Bright
In a more golden light
I see
Each little doubt and fear,
Each little discord here
Removed.

Then will I not repine,
Knowing that Bird of mine—
Though flown—
Shall in a distant tree
Bright melody for me
Return.

I dreaded that first Robin so,
But He is mastered now,
And I'm accustomed to Him grown,
He hurts a little, though—

I thought if I could only live
Till that first Shout got by—
Not all Pianos in the Woods
Had power to mangle me—

I dared not meet the Daffodils,
For fear their Yellow Gown
Would pierce me with a fashion
So foreign to my own—

I wished the Grass would hurry—
So—when 'twas time to see—
He'd be too tall, the tallest one
Could stretch—to look at me—

I could not bear the Bees should come,
I wished they'd stay away
In those dim countries where they go,
What word had they, for me?

They're here, though; not a creature failed—
No Blossom stayed away
In gentle deference to me—
The Queen of Calvary—

Each one salutes me as he goes,
And I my childish Plumes,
Lift, in bereaved acknowledgment
Of their unthinking Drums—

Love's Good-Morrow

Pack, clouds away! and welcome day!
    With night we banish sorrow;
Sweet air, blow soft, mount larks aloft
    To give my love good-morrow!
Wings from the wind to please her mind,
    Notes from the lark I'll borrow;
Bird, prune thy wing, nightingale, sing,
    To give my love good-morrow;
    To give my love good-morrow;
    Notes from them both I'll borrow.

Wake from thy nest, Robin Redbreast,
    Sing birds in every furrow;
And from each hill, let music shrill
    Give my fair love good-morrow!
Blackbird and thrush in every bush,
    Stare, linnet, and cock-sparrow!
You pretty elves, amongst yourselves,
    Sing my fair love good-morrow;
    To give my love good-morrow,
    Sing birds in every furrow.

THOMAS HEYWOOD

To Robin Red-Breast

Laid out for dead, let thy last kindness be
With leaves and moss-work for to cover me;
And while the wood-nymphs my cold corpse inter,
Sing thou my dirge, sweet-warbling chorister!
For epitaph, in foliage, next write this:
HERE, HERE THE TOMB OF ROBIN HERRICK IS!

ROBERT HERRICK

Poor Cock Robin

My garden robin in the Spring
Was rapturous with glee,
And followed me with wistful wing
From pear to apple tree;
His melodies the summer long
He carolled with delight,
As if he could with jewelled song
Find favour in my sight.

And now that Autumn's in the air
He's singing singing still,
And yet somehow I cannot bear
The frenzy of his bill;
The keen wind ruffs his ruddy breast
As to bare boughs he clings;
The sun is sullen in the West
Yet still he sings and sings.

Soon, soon the legions of the snow
Will pitch their tents again,
And round my window-sill I know
He'll call for crumbs in vein;
The pulsing passion of his throat
Has hint of Winter woe;
The piercing sweetness of his note
entreats me not to go.

In vein, in vain, Oh valiant one,
You sing to bid me stay!
For all my life is in the sun
And I must fly away.
yet by no gold or orange glow
Will I be comforted,
Seeing blood-bright in bitter snow—
              A robin dead.

DAS 2004

European Starling

The first hundred European Starlings in North America were released in New York's Central Park in 1890 as part of an effort to introduce into the United States every species of bird mentioned in Shakespeare's works. Now it is one of the most widespread birds on the continent and is thought to have displaced some native species.

The Death of Lesbia's Bird

Pity! mourn in plaintive tone
The lovely starling dead and gone!
Pity mourns in plaintive tone
The lovely starling dead and gone.
Weep, ye Loves! and Venus! weep
The lovely starling fallen asleep!
Venus sees with tearful eyes—
In her lap the starling lies!
While the Loves all in a ring
Softly stroke the stiffened wing.

Translated by Samuel Taylor Coleridge

Tonight, in the country, I stood awhile
under the dying oaks—the caterpillars
eat everything but stone—and one by one
starlings flung themselves at me
from the trees, pulling up
at the last moment and sweeping back
into the ornamentally fading sky.
At first I thought they were simply
playful, but they got closer and closer,
their whistling shrill.

If just one bird flew with intent into the window,
shattering the glass, I could believe
that there is something of substance
in their lives, but their bones are hollow—
I imagine that
without marrow they can't suffer,
that misery won't exist in air:
once torn from the earth
it is lifted up and dispersed.

It is still light out and I return
to prepare dinner. I have
something sorrowful keeping me here.

Somewhere there are people beginning
to make fires. Their lives,
of earthly marrow, work the fields.
At night, when the sun sets but it's still
early, women prepare small birds,
unstuffed, bony,
and the family cats.
And then it sleeps.
One of the young boys or girls
enters a weightless dream and rises,
not in the night, but through the air.

Cedar Waxwing

Although it catches insects during the breeding season, the Cedar Waxwing eats primarily fruit—unusual for a bird whose range is in temperate rather than tropical habitats—and can live on fruit alone for several months. As the poem indicates, it is vulnerable to alcohol intoxication and can die after eating fermented fruit. It is found throughout North America.

Cedar Waxwings

A dozen of them dodged and fluttered
in the branches of the thirsty rhododendron
being drenched by our backyard sprinkler. Some perched
among the leaves holding their wings open to the water
as others, a little apart, shrugged themselves dry.

I lost count as more kept arriving
in their black burglar masks, brown or black
throat scarves, olive green jackets and crested hats,
yellow trim at the end of their tails. Those in command
flaunted bright red flashing near their wingtips.

What could have led people in past times, I wondered,
to regard these birds as harbingers of death?
They're tame and sociable. They call to each other in flight.
Several may sit together on a branch
or wire, passing a piece of fruit back and forth

beak to beak, sharing the taste. Mating pairs do this
with flower petals. An adult can hold
as many as thirty chokecherries in its crop
and regurgitate them one by one into the mouths of its young.
They love to party. Sometimes they get so drunk

on overripe berries they keel over
and then have to sleep it off.
The branches they flocked on bobbed and sagged, and the air
was full of their gleeful gibberish.
Not one of them weighed more than an ounce.

JONATHAN AARON

Cedar Waxwing on Scarlet Firethorn

To start again with something beautiful,
and natural, the waxwing first on one
foot, then the other, holding the berry
against the moment like a drop of blood—
red-wing-tipped, yellow at the tip of the
tail, the head sleek, crested, fin or arrow,
turning now, swallowing. Or any bird
that turns, as by instruction, its small, dark
head, disinterested, toward the future; flies
into the massive tangle of the trees, slick.
The visual glide of the detail blurs.

The good gun flowering in the mouth is done,
like swallowing the sword or eating fire,
the carnival trick we could take back if
we wanted. When I was told suicide
meant the soul stayed with the body locked in
the ground I knew it was wrong, that each bird
could be anyone in the afterlife,
alive, on wing. Like this one, which lets its
thin lisp of a song go out into the wood-
land understory, into its voice, gone.

But to look down the long shaft of the air,
The whole healing silence of the air, fire
and thorn, where we want to be, on the edge
of the advantage, the abrupt green edge
between the flowering pyracantha and
the winded, open field, before the trees—
to be alive in secret, this is what
we wanted, and here, as when we die what
lives is fluted on the air—a whistle,
then the wing—even our desire to die,
to swallow fire, disappear, be nothing.

The body fills with light, and in the mind
the white oak of the table, the ladder
stiffness of the chair, the dried-out paper
on the wall fly back into the vein and
branching of the leaf—flare like the waxwings,
whose moment seems to fill the scarlet hedge.
From the window, at a distance, just more
trees against the sky, and in the distance
after that everything is possible.
We are in a room with all the loved ones,
who, when they answer, have the power of song.

* * * * * * * * * * * * * * * * * * *

STANLEY PLUMLY

DAS 2004

Blackburnian Warbler

A bird of the coniferous and mixed forests of the northeastern United States, the Blackburnian Warbler is the only North American warbler to have an orange throat. Although New World warblers can be quite colorful, Old World warblers are drab— it is likely that Keats's "ruby-breasted warbler" was a European Robin.

Song

Stay, ruby-breasted warbler, stay,
    And let me see thy sparkling eye,
Oh brush not yet the pearl-strung spray
    Nor bow thy pretty head to fly.

Stay while I tell thee, fluttering thing,
    That thou of love an emblem art,
Yes! patient plume thy little wing,
    Whilst I my thoughts to thee impart.

When summer nights the dews bestow,
    And summer suns enrich the day,
Thy notes the blossoms charm to blow,
    Each opes delighted at thy lay.

So when in youth the eye's dark glance
    Speaks pleasure from its circle bright,
The tones of love our joys enhance
    And make superior each delight.

And when bleak storms resistless rove,
    And every rural bliss destroy,
Nought comforts then the leafless grove
    But thy soft note—its only joy—

E'en so the words of love beguile
    When Pleasure's tree no longer bears,
And draw a soft endearing smile
    Amid the gloom of grief and tears.

. . . .

JOHN KEATS

A Warbler Announcing Spring at Morning

How does he know
that spring has come to the world?
From
        within the cage
where he wakes from sleep at daybreak—
the sound of a warbler's call.

Translated by Stephen D. Carter

SHŌTETSU

The Happy Bird

The happy whitethroat on the sweeing bough,
Swayed by the impulse of the gadding wind
That ushers in the showers of April, now
Carols right joyously; and now reclined,
Crouching, she clings close to her moving seat,
To keep her hold; and till the wind for rest
Pauses, she mutters inward melodies,
That seem her heart's rich thinkings to repeat.
But when the branch is still, her little breast
Swells out in rapture's gushing symphonies;
And then, against her brown wing softly prest,
The wind comes playing, an enraptured guest,
This way and that she swees—till gusts arise
More boisterous in their play, then off she flies.

JOHN CLARE

Yellow Warbler

The Yellow Warbler, found in deciduous thickets and in mangrove swamps throughout North America, can be distinguished from other yellow birds by the red streaks on the chest of the male. It is the only warbler with yellow tail spots.

Every time we put crumbs out and sunflower
seeds something comes. Most often sparrows.
Frequently a jay. Now and then a junco or
a cardinal. And once—immediately and never
again, but as commonly as any miracle while it
is happening, and then instantly incredible for-
ever—the tiniest (was it?) yellow warbler
as nearly as I could thumb through the bird
book for it, or was it an escaped canary? or
simply the one impossible bright bird that is
always there during a miracle, and then never?

I, certainly, do not know all that comes to us
at times. A bird is a bird as long as it is
there. Then it is a miracle our crumbs and
sunflower seeds caught and let go. Is there
a book to look through for the identity
of a miracle? No bird that is there is
miracle enough. Every bird that has been is
entirely one. And if some miracles are rarer
than others, every incredible bird has crumbs
and seeds in common with every other. Let there
be bread and seed in time: all else will follow.

DAS 2006

Northern Cardinal

The Northern Cardinal, a large long-tailed finch, may be the most popular
bird in the United States, as no fewer than seven states claim it as their state bird.
It does not migrate, and the male defends his territory from other cardinals
quite fiercely. The female sings, which is rare among female North American
songbirds. It is abundant across the eastern United States, from Maine to Texas,
and in some parts of Canada.

The Cardinal

Not to conform to any other color
is the secret of being colorful.

He shocks us when he flies
like a red verb over the snow.

He sifts through the blue evenings
to his roost.

He is turning purple.
Soon he'll be black.

In the bar's dark I think of him.
There are no cardinals here.

Only a woman in a red dress.

Cardinals in a Shower at Union Square

At first they look like any other birds
on gun line from the underbrush, so someone
calls them sparrows and someone who thinks
he knows, scarlet tanagers or something else
exotic, as if they've slipped captivity—
one of those white sky August days the hammer
of the heat picks out the old one or the child
locked in a car, while gathered above the blank
grave of the pavement, at the altitude of snow,
enough rain to almost forgive it all.
Only two are really red, the rest a buoyant
dried blood brown, young or female, all of them
with masks and crests that make them what
they are, explosions from the other side
or blown in, with the paper, with the storm.
Whoever starts the clapping is answered
by a show of hands to meet baptismal waters
and a couple, who are high, bird-dancing.
Whoever starts the shouting is quieted
by the lady who hears silences,
cupping her clownish ears. . . .
For a moment the ringing air is clean, then
for a moment nothing happens, nothing moves
except the cardinals, in and out of trees,

And in that moment ends. The cloudburst
passes, the air turns into fire again,
the sirens sing their distances, the walls
of light burn down. And in no time,
in the time it takes the runoff to drain
back underground, there's no one left
but lifers and the dealers and rain birds
swallowed upward by the sun, and rain, new rain,
in the rivers and the reservoir uptown,
ready to rise and pour its heart out all over.

Western Tanager

The pair described in the poem are Scarlet Tanagers, which have been affected by forest fragmentation in the eastern United States. The Western Tanager, a resident of coniferous and mixed forests, does not face the same threat of habitat loss. It is found across the western United States and Canada, from the Mexican border to Alaska.

A Pair of Tanagers

The scarlet male, his green mate, their black wings
Beside the A/C unit in the dull dirt:

They look at first like a child's abandoned toys.
But ants and iridescent flies have found them,

Working along the seams of the shut beaks
And the dark indentations of the eyelids.

You want to give something like this a moral:
Like, the woods these days are full of hard illusions,

Or, never fly north if you think you're flying south,
Or, stay above rooftops; if you meet yourself

Coming, it's too late; death is a big surprise.
And their death together certainly startles us.

Stopped short. But how recently in the rain forest,
How recently in the place they were first named,

Reflected on the Amazon, the Orinoco,
Headlong from Brazil, into our window.

You want to give something like this a moral
Or see it as an omen, as a portent.

And then, the long journeying comes to mind,
Together such a distance, to this end.

MARK JARMAN

DAS 2005

Chipping Sparrow

The smallest sparrow, the Chipping Sparrow is fairly tame. Originally a bird of the coniferous forest, it has adapted well to modifications of its habitat and is commonly seen in gardens, farmlands, and forest clearings. It is common across North America.

whose home is in the straw
and bailing twine threaded
in the slots of a roof vent

who guards a tiny ledge
against the starlings
that cruise the neighborhood

whose heart is smaller
than a heart should be,
whose feathers stiffen

like an arrow fret to quicken
the hydraulics of its wings,
stay there on the metal

ledge, widen your alarming
beak, but do not flee as others have
to the black walnut vaulting

overhead. Do not move outside
the world you've made
from bailing twine and straw.

The isolated starling fears
the crows, the crows gang up
to rout a hawk. The hawk

is cold. And cold is what
a larger heart maintains.
The owl at dusk and dawn,

far off, unseen, but audible,
repeats its syncopated intervals,
a song that's not a cry

but a whisper rising from concentric
rings of water spreading out across
the surface of a catchment pond.

It asks, "Who are you? Who
are you?" but no one knows.
Stay where you are, nervous, jittery.

Move your small head a hundred
ways, a hundred times, keep
paying attention to the terrifying

world. And if you see the Robins
in their dirty orange vests
patrolling the yard like thugs,

forget about the worm. Starve
yourself, or from the air inhale
the water you may need, digest

the dust. And what the promiscuous
cat and jaybirds do, let them
do it, let them dart and snipe,

let them sound like others.
They sleep when the owl sends
out its encircling question.

Stay where you are, you lit fuse,
you dull spark of saltpeter and sulfur.

. . . .

MICHAEL COLLIER

Love Song

Love comes hungry to anyone's hand.
I found the newborn sparrow next to
the tumbled nest on the grass. Bravely

opening its beak. Cats circled, squirrels.
I tried to set the nest right but the wild
birds had fled. The knot of pinfeathers

sat in my hand and spoke. Just because
I've raised it by touch, doesn't mean it
follows. All day it pecks at the tin image of

a faceless bird. It refuses to fly,
though I've opened the door. What
sends us to each other? He and I

had a blue landscape, a village street,
some poems, bread on a plate. Love
was a camera in a doorway, love was

a script, a tin bird. Love was faceless,
even when we'd memorized each other's
lines. Love was hungry, love was faceless,

the sparrow sings, famished, in my hand.

CAROL MUSKE-DUKES

Christmas Sparrow

The first thing I heard this morning
was a soft, insistent rustle,
the rapid flapping of wings
against glass as it turned out,

a small bird rioting
in the frame of a high window,
trying to hurl itself through
the enigma of transparency into the spacious light.

A noise in the throat of the cat
hunkered on the rug
told me how the bird had gotten inside,
carried in the cold night
through the flap in a basement door,
and later released from the soft clench of teeth.

Up on a chair, I trapped its pulsations
in a small towel and carried it to the door,
so weightless it seemed
to have vanished into the nest of cloth.

But outside, it burst
from my uncupped hands into its element,
dipping over the dormant garden
in a spasm of wingbeats
and disappearing over a tall row of hemlocks.

Still, for the rest of the day,
I could feel its wild thrumming
against my palms whenever I thought
about the hours the bird must have spent
pent in the shadows of that room,
hidden in the spiky branches
of our decorated tree, breathing there
among the metallic angels, ceramic apples, stars of yarn,

its eyes open, like mine as I lie here tonight
picturing this rare, lucky sparrow
tucked into a holly bush now,
a light snow tumbling through the windless dark.

BILLY COLLINS

House Sparrow

A native of Europe and parts of Asia, the House Sparrow was introduced into the United States in Brooklyn, New York, in 1851 and had spread to the Rocky Mountains by 1900. It has been in both North and Central America long enough for evolution to have shaped it: populations in the north are larger than those in the south, as is generally true with native species.

Sparrow Trapped in the Airport

Never the bark and abalone mask
cracked by storms of a mastering god,
never the gods' favored glamour, never
the pelagic messenger bearing orchards
in its beak, never allegory, not wisdom
or valor or cunning, much less hunger
demanding vigilance, Industry, Invention,
or the Instinct to claim some small rise
above the plain and from there to assert
the song of another day ending;
lentil brown, uncounted, overlooked
In the clamorous public of the flock
so unlikely to be noticed here by arrivals,
faces shining with oils of their many miles,
where it hops and scratches below
the baggage carousel and lights too high,
too bright for any real illumination,
looking more like a fumbled punch line
than a stowaway whose revelation
recalls how lightly we once traveled.

Partita for Sparrows

We bury the sparrows of Europe
with found instruments,
their breasts light as an ounce of tea
where we had seen them off the path,
their twin speeds of shyness and notched wings
near the pawnbroker's house by the canal,
in average neighborhoods of the resisters,
or in markets of princely delphinium and flax,
flying from awnings at unmarked rates
to fetch crumbs from our table half-spinning
back to clefs of grillwork on external stairs
we would descend much later;

in rainy neighborhoods of the resisters
where streets were taken one by one,
where consciousness is a stair or path,
we mark their domains with notched sticks
of hickory or chestnut or ash
because our cities of princely pallor
should not have unmarked graves.
Lyric work, flight of arch, death bridge
to which patterned being is parallel:
they came as if from the margins
of a painting, their average hearts half-spinning
our little hourglass up on the screen.

BRENDA HILLMAN

Walking past the open window, she is surprised
by the song of the white-throated sparrow
and stops to listen. She has been thinking of
the dead ones she loves—her father who lived
over a century, and her oldest son, suddenly gone
at forty-seven—and she can't help thinking
she has called them back, that they are calling her
In the voices of these birds passing through Ohio
on their spring migration . . . because, after years
of summers in upstate New York, the white-throat
has become something like the family bird.
Her father used to stop whatever he was doing
and point out its clear, whistling song. She hears it
again: "Poor Sam Peabody Peabody Peabody."
She tries not to think, "Poor Andy," but she
has already thought it, and now she is weeping.
But then she hears another, so clear, it's as if
the bird were in the room with her, or in her head,
telling her that everything will be all right.
She cannot see them from her second story window—
they are hidden in the new leaves of the old maple,
or behind the white blossoms of the dogwood—
but she stands and listens, knowing they will stay
for only a few days before moving on.

JEFFREY HARRISON

White throated sparrow, the century is beginning.
   The years ahead are waiting like white seeds.
      History is your silvery evening call,
         Whistling the name of another place you love
            As you return, then go back into hiding,
               The browns and brights of winter in your feathers.

Small migrator, we watch as you ignore
   History altogether, all that is not seeds,
      In synch with seasons but not any event
         That stops a human being with a thought
            (Though this event has stopped me). Your wings flex
               Like shadows of the ice cap and the leaves.

You are as you have always been, we hope.
   If one of us moves too quickly, you depart.
      You travel in neat kin groups over latitudes.
         You prefer to eat the seeds spread on the ground.
            The centuries have changed, there are new numbers.
               We are counting on many things, including you.

MARK JARMAN

DAS 2003

Dark-eyed Junco

A widespread small sparrow, the Dark-eyed Junco can be seen as easily in
New York as in Texas. There are seven subspecies, distinguishable by their coloring
and patterning, that inhabit different areas of Canada, the United States, and
northern Mexico. Its song is described as a musical trill, with short hard
smack and *tick* calls.

The Call of the Junco Bird

An English woman I've never met
calls to read me her new poem
about the little Texas junco bird
whose cry sounded to the early settlers
like the words, *no hope, no hope.*

The bird knows what it has taken her
half a lifetime to learn,
she says, now that her body
is covered with sores
and she can no longer walk.

She needs me to go to the yard
and listen to the desolate plea
of a bird I've never seen,
a song I've never heard: hope
is no longer a thing with feathers.

Night pauses with its ear cocked·
Listen to the cry of the female,
she calls, who has drunk herself
into a stupor and trills in high C
the words, *no hope, no hope.*

DAS 2003

Red-winged Blackbird

The Red-winged Blackbird, a medium-size songbird, is perhaps the most
abundant bird in North America. It is a highly polygynous species, with up to
fifteen females making nests in the territory of one male. The brown, striped
female could be taken for a different species than the black male, with his
distinctive red-orange epaulets.

Redwing Blackbirds

How far a-winging to keep this appointment with April!
How much breath left in reserve to fill
The sky of washed azure and whipped-cream cumuli
With their rusty, musical, heart-plumbing cry!

On sedge, winter-bit but erect, on old cattails, they swing.
Throats throb, your field glasses say, as they cling and sing—
If singing is what you call that rusty, gut-grabbing cry
That calls on life to be lived gladly, gladly.

They twist, tumble, tangle, they glide and curvet,
And sun stabs the red splash to scarlet on each epaulet.
And the lazy distance of hills seems to take
A glint more green, and dry grass at your feet to wake.

In the vast of night, seasons later, sleet coding on pane,
Fire dead on hearth, hope banked in heart, I again
Awake, not in dream but with eyes shut, believing I hear
That rusty music far off, far off, and catch flash and fleer

Of a scarlet slash accenting the glossy black. Sleet
Continues. The heart continues its steady beat
As I burrow into the tumulus of sleep,
Where all things are buried, though no man for sure knows how deep.

The globe grinds on, proceeds with the business of Aprils and men.
Next year will redwings see me, or I them, again then?
If not, some man else may pause, awaiting that rusty, musical cry,
And catch—how gallant—the flash of epaulets scarlet against blue sky.

ROBERT PENN WARREN

Red-Winged Blackbirds

The epaulettes redeem them; otherwise
I'd hate them, corvids stealing seed I'd bought
for cardinals and sparrows. But there's that splash
of color: a swatch of yellow on the wing,
a hint of scarlet underneath that makes
their other feathers blacker than any raven's.
They're pretty, so I like them. I know it's just
the males who wear such fancy duds, and no,
I haven't missed the irony of that.
Here's the point where I should turn the birds
to metaphors, embodiments of sex
or beauty, nature's cunning artifice,
a broken heart or other human flaw.
Their shoulders blaze like eyes, like coals, like wounds,
like circumstance as they stretch and fly away.

JULIANA GRAY

DAS 2005

Tricolored Blackbird

Closely related to the Red-winged Blackbird, the Tricolored Blackbird lives only in central and southern California and northern Baja California, with small breeding colonies in western Nevada, Oregon, and southern Washington.

The Birds

The world begins again!
Not wholly insufflated
the blackbirds in the rain
upon the dead topbranches
of the living tree,
stuck fast to the low clouds,
notate the dawn.
Their shrill cries sound
announcing appetite
and drop among the bending roses
and the dripping grass.

WILLIAM CARLOS WILLIAMS

Thirteen Ways of Looking at a Blackbird

I

Among twenty snowy mountains,
The only moving thing
Was the eye of the blackbird.

II

I was of three minds,
Like a tree
In which there are three blackbirds.

III

The blackbird whirled in the autumn winds.
It was a small part of the pantomime.

IV

A man and a woman
Are one.
A man and a woman and a blackbird
Are one.

V

I do not know which to prefer,
The beauty of inflections
Or the beauty of innuendoes,
The blackbird whistling
Or just after.

VI

Icicles filled the long window
With barbaric glass.
The shadow of the blackbird
Crossed it, to and fro.
The mood
Traced in the shadow
An indecipherable cause.

VII

O thin men of Haddam,
Why do you imagine golden birds?
Do you not see how the blackbird
Walks around the feet
Of the women about you?

. . . .

VIII

I know noble accents
And lucid, inescapable rhythms;
But I know, too,
That the blackbird is involved
In what I know.

IX

When the blackbird flew out of sight,
It marked the edge
Of one of many circles.

X

At the sight of blackbirds
Flying in a green light,
Even the bawds of euphony
Would cry out sharply.

XI

He rode over Connecticut
In a glass coach.
Once, a fear pierced him,
In that he mistook
The shadow of his equipage
For blackbirds.

XII

The river is moving.
The blackbird must be flying.

XIII

It was evening all afternoon.
It was snowing
And it was going to snow.
The blackbird sat
In the cedar-limbs.

. . . .

DAS 2005

Common Grackle

An opportunistic forager on suburban lawns, on plowed fields, and in shallow
water, the Common Grackle can be recognized by its iridescent purple and bronze
feathers. Possibly to rid itself of parasites, it allows ants to crawl on its feathers
and release formic acid, but also may employ lemons and limes, chokecherries, and
even mothballs for the same purpose. It is common in North America east
of the Rocky Mountains.

One grackle two grackles in the maple three four
two grackles one grackle in the maple none grackles
in the maple: I do ask them something
by looking at them, as they ask me nothing
by not looking at me, what is sky to birds,
four grackles in the maple and this sense of sky
in my head. As soon as five six seven eight
arrive, four hops to a different branch,
then a shuttling in and out like our subwaying
to work, now one, now none grackles, I am poor.
Then it is later and still none grackles,
still alone, though behind me, where I can't see,
some tweet, chirp, what am I, a xylophone?
I translate best I can, now a gaggle of, a swarm of
six, who knows for sure what singings
they really are? I don't, my ornithology's weak,
and while confessing my poor birding, six left,
the air favors minus slightly more than plus,
though I can look the air in the eye
and hold what is to what was. Was grackles,
was some need I had to feel
mending going on, without knowing
what's skewed or rent, now a crow
making the tree resemble an excuse for crow,
as I am an excuse for death to take its time.

BOB HICOK

They were not one body. Yet they seemed
held together by some order, their thick necks
flickering with a blue-black iridescence,
their yellow-circled pupils bright and cold.

In a wave of differences that passed
low over the surface of my yard,
they picked it clean of morning's fritillaries
and other summer gestures fall discards

then settled on the hill behind the fence
for several teeming minutes to remark
its tapestry, each razored beak, each tail
parting Sunday's gray air like a spear.

I could tell you that they gathered up
the darkness of my winter thought that day
in mid-September, bundled it, black-ribboned,
into sleek coats and lifted it from me

just as you have imagined. But this
would be a lie. I watched them comb the fields
with interest, and, when their beak's clicks had died,
turned back to what I was.

LISA WILLIAMS

Rose-breasted Grosbeak

The male Rose-breasted Grosbeak—a striking black-and-white bird with a vivid red chest marking, as opposed to the drab, brown-and-white-striped female—participates in the incubation of the eggs, spending about one-third of his time on the nest during the day. The birds sing to each other as they exchange places. Its range is across most of Canada and the eastern United States.

Rose-Breasted Grosbeak

Oh, pretty bird! Oh, fluff and feathers, beak
and bright eye, alliterative name
in my throat! Too easy reach to say *lover*,
heart's blood, blush or *flush* or *crush of wine-*
dark berries over snow. Too easy now
to smile at window glass. Pretty boy,
misjudging you was never just a game.

American Goldfinch

The American Goldfinch is especially fond of thistles, eating the seeds and lining its nest with the down. It changes from breeding plumage—lemon-yellow body and black forehead, cap, wings, and tail for the male, and yellowish-brown body and blackish-brown wings and tail for the female—to winter plumage by a complete molt of all its feathers. The male is mostly monogamous, but some females change mates after the first brood; the first male takes care of the fledglings while the female starts a new brood with a different male. It is found throughout much of North America.

The Faithful Friend

The green-house is my summer seat;
My shrubs displac'd from that retreat
Enjoy'd the open air;
Two goldfinches, whose sprightly song
Had been their mutual solace long,
Liv'd happy pris'ners there.

They sang, as blithe as finches sing
That flutter loose on golden wing,
And frolic where they list;
Strangers to liberty, 'tis true,
But that delight they never knew,
And, therefore, never miss'd.

But nature works in ev'ry breast;
Instinct is never quite suppress'd;
And Dick felt some desires,
Which, after many an effort vain,
Instructed him at length to gain
A pass between his wires.

The open windows seem'd to invite
The freeman to a farewell flight;
But Tom was still confin'd;
And Dick, although his way was clear,
Was much too gen'rous and sincere
To leave his friend behind.

For, settling on his grated roof,
He chirp'd and kiss'd him, giving proof
That he desir'd no more;
Nor would forsake his cage at last,
Till gently seiz'd I shut him fast,
A pris'ner as before.

Oh ye, who never knew the joys
Of Friendship, satisfied with noise,
Fandango, ball and rout!
Blush, when I tell you how a bird,
A prison, with a friend, preferr'd
To liberty without.

WILLIAM COWPER

The Caged Goldfinch

Within a churchyard, on a recent grave,
I saw a little cage
That jailed a goldfinch. All was silence save
Its hops from stage to stage.

There was inquiry in its wistful eye,
And once it tried to sing;
Of him or her who placed it there, and why,
No one knew anything.

True, a woman was found drowned the day ensuing,
And some at times averred
The grave to be her false one's, who when wooing
Gave her the bird.

For the Birds: A Charm of Goldfinches

Stopped under a sycamore, looked up:
bare white limbs against blue, blue sky
and in those branches, flickering, birds,
each with a pale green-yellow breast,
each the size of a small child's fist.
What kind of birds are you? I asked

and put on my glasses, the better to glimpse
such wing and color, such flashiness.
Then breathless climbed the sun-swept hill
to the visitor's center, rushed inside,
saying, "I have a question about a bird!"
Was given a book of birds to check.

Considered *Common Yellow Throat:*
Skulks in marshes.
Male wears black mask.
Wichity-wichity song
Loved that music, but maybe I'm wrong?
Was told that American Goldfinches turn

from winter's muddy greenish-brown
to summer's yellow brightness, turn
betwixt, in spring, this lemon-lime;
and fly *in hiccups*, flash their gold, a flock
of such birds called *a charm*,
from the Latin *carmen*, meaning *song*.

Ran back down the hill like a woman afire
practically into the sycamore's arms,
singing, anyway, *skulks in marshes*,
black mask, wichity-wichity song!
Singing, *Spread out your colors, oh flash me your wings*
as the charm made its green-yellow sweep through the sky.

When interviewed, the bird watchers were tight-
lipped. One actually got up and left.
When interviewed, the bird watchers gave
quick, birdlike answers. We had to ask many
of our questions twice.
When interviewed, the bird watchers wore hoods.
When interviewed, the bird watchers had this
to say, *We are no more boring than scientists
or stonemasons. Indeed, many of us are scientists
and stonemasons. Or we could be.*
When interviewed, the bird watchers ate crow.
When interviewed, the bird watchers refused
to discuss themselves, preferring to debate
recent observations of the Dope Warbler,
the Spoon Tailed Ninny, the Royal Bavarian
Snack Rail, whether any would survive
another New Hampshire winter.
When interviewed, the bird watchers sucked eggs.
When interviewed, the bird watchers placed
blame squarely on the coal industry.
When interviewed, the bird watchers turned
on each other. It was ugly, like a stoning.
When interviewed, the bird watchers began
to undress, slowly, as if half asleep.

When interviewed, the bird watchers freaked;
one threw handfuls of dirt at the cameras,
another wept uncontrollably.
When interviewed, the bird watchers were small.
When interviewed, the bird watchers blushed
like startled lovers, like priests and nuns.
When interviewed, the bird watchers varied
in their appraisal of the year's display. One
called it *bounteous, rich*. But another seemed
confused, described the birds as *cumulous*,
the weather as *sinusoidal*.
When interviewed, the bird watchers appeared
to be holding back, hiding something.
When interviewed, the bird watchers agreed
to be interviewed again.
The birds could not be reached for comment.

BRENDAN CONSTANTINE

PERMISSIONS

. . . .

• • • • • • • • • • • • • • • • • • •

Halpern, Daniel. "Of the Air" from *Seasonal Rights* by Daniel Halpern, copyright © 1979, 1980, 1981, 1982 by Daniel Halpern. Used by permission of Viking Penguin, a division of Penguin Group (USA) Inc.

Harrison, Jeffrey. "Visitation" from *Incomplete Knowledge* by Jeffrey Harrison. Copyright © 2006 by Four Way Books.

Heaney, Seamus. "Drifting Off" from *Opened Ground: Selected Poems, 1966–1996*, by Seamus Heaney. Copyright © 1998 by Seamus Heaney. Reprinted by permission of Farrar, Straus and Giroux, LLC, and Faber and Faber Ltd.

Hicok, Bob. "Keeping track" used by kind permission of the author.

Hillman, Brenda. "Partita for Sparrows" first published in *The Modern Review*. Used by kind permission of the author.

Hirsch, Edward. "The Call of the Junco Bird" first published in *The Nation*. Used by kind permission of the author.

Hirshfield, Jane. "Hope and Love" from *The Lives of the Heart* by Jane Hirshfield. Copyright © 1997 by Jane Hirshfield. Reprinted by permission of HarperCollins Publishers and Michael Katz. "The Woodpecker Keeps Returning" from *After: Poems* by Jane Hirshfield. Copyright © 2006 by Jane Hirshfield. Reprinted by permission of HarperCollins Publishers and Michael Katz.

Hollander, John. "Swan and Shadow" from *Types of Shape* by John Hollander. Copyright © 1991 by Yale University Press.

Jamie, Kathleen. "Plover" by Kathleen Jamie. Copyright © Kathleen Jamie. Reproduced by permission of the author c/o Rogers, Coleridge & White Ltd., 20 Powis Mews, London W11 1JN.

Jarman, Mark. "Old Acquaintance" and "A Pair of Tanagers" from *To the Green Man*. Copyright © 2004 by Mark Jarman. Reprinted with the permission of Sarabande Books, www.sarabandebooks.org. "Chimney Swifts" reprinted with the kind permission of the author.

Laux, Dorianne. "The Ravens of Denali" from *Facts About the Moon* by Dorianne Laux. Copyright © 2006 by Dorianne Laux. Used by permission of W. W. Norton & Company, Inc.

McCleery, Nancy. "December Notes" from *Girl Talk* copyright © 2002 by Nancy McCleery. Used by kind permission of the author.

Mehigan, Joshua. "A Bird at the Leather Mill" from *The Optimist: Poems* by Joshua Mehigan. Copyright © 2004. Reprinted with the permission of Ohio University Press/Swallow Press, Athens, Ohio (www.ohioswallow.com).

Mitchell, Susan. "Lost Parrot" first printed in *The Atlantic* and used with the kind permission of the author.

Moore, Marianne. "The Frigate Pelican" reprinted with the permission of Scribner, a Division of Simon & Schuster, Inc., from *The Collected Poems of Marianne Moore* by Marianne Moore. Copyright © 1935 by Marianne Moore. Copyright renewed © 1963 by Marianne Moore and T. S. Eliot. All rights reserved. Reprinted in the UK, Ireland, and Commonwealth with the permission of Faber and Faber Ltd.

Muske-Dukes, Carol. "Love Song" from *Sparrow* by Carol Muske-Dukes, copyright © 2003 by Carol Muske-Dukes. Used by permission of Random House, Inc.

Nemerov, Howard. "The Blue Swallows" from *The Blue Swallows* copyright © 1967 by Howard Nemerov. Used by kind permission of Margaret Nemerov.

• • • • • • • • • • • • • • • • • • •

Schwartz, Delmore. "The Ballet of the Fifth Year" from *Selected Poems: Summer Knowledge* by Delmore Schwartz, copyright © 1938 by Delmore Schwartz. Reprinted by permission of New Directions Publishing Corp. Reproduced by permission of Pollinger Limited in the UK, Ireland, and the Commonwealth.

Schwartz, Ruth L. "The Swan at Edgewater Park" from *Edgewater* by Ruth L. Schwartz. Copyright © 2002 by Ruth L. Schwartz. Reprinted by permission of HarperCollins Publishers.

Simic, Charles. "The Great Horned Owl" from *Selected Early Poems* by Charles Simic. Copyright © 1999. Reprinted with permission of George Braziller, New York.

Snyder, Gary. "Magpie's Song" from *Turtle Island* by Gary Snyder. Copyright © 1974 by Gary Snyder. Reprinted with permission of New Directions Publishing Group.

Steele, Timothy. "Black Phoebe" from *Toward the Winter Solstice: New Poems* by Timothy Steele. Copyright © 2006. Reprinted with the permission of Ohio University Press/Swallow Press, Athens, Ohio (www.ohioswallow.com).

Stein, Kevin. "Arts of Joy" from *Sufficiency of the Actual* by Kevin Stein. Copyright © 2008. Reprinted with permission of University of Illinois Press.

Stern, Gerald. "Swifts" from *Last Blue: Poems* by Gerald Stern. Copyright © 2000 by Gerald Stern. Used by permission of W. W. Norton & Company, Inc.

Stevens, Wallace. "Thirteen Ways of Looking at a Blackbird" from *The Collected Poems of Wallace Stevens* by Wallace Stevens, copyright © 1954 by Wallace Stevens and renewed 1982 by Holly Stevens. Used by permission of Alfred A. Knopf, a division of Random House, Inc.

Stevenson, Anne. "Swifts" by Anne Stevenson from *Poems, 1955–2005*. Copyright © 2005. Reprinted with permission of Bloodaxe Books.

• • • • • • • • • • • • • • • • • • • •

Tate, James. "The Blue Booby" from *Selected Poems* © 1991 by James Tate and reprinted by permission of Wesleyan University Press.

Updike, John. "Seagulls" from *Collected Poems, 1953–1993*, by John Updike, copyright © 1993 by John Updike. Used by permission of Alfred A. Knopf, a division of Random House, Inc.

Wagoner, David. "Loons Mating" from *Traveling Light: Collected and New Poems*. Copyright 1999 by David Wagoner. Used with permission of the poet and the University of Illinois Press. "To a Farmer Who Hung Five Hawks on His Barbed Wire" from *First Light* by David Wagoner. Copyright © 1983 by David Wagoner. By permission of Little, Brown & Company.

Warren, Robert Penn. "Evening Hawk" and "Redwing Blackbirds" are from *The Collected Poems of Robert Penn Warren* Copyright © 1998 by Estate of Robert Penn Warren. Reprinted by permission of William Morris Agency, LLC, on behalf of the Author.

Wilbur, Richard. "Still, Citizen Sparrow" from *Ceremony and Other Poems* © 1965 by Richard Wilbur and reprinted with the permission of Houghton Mifflin Harcourt Publishing Company. "A Barred Owl" from *Mayflies* © 2000 by Richard Wilbur and reprinted with the permission of Houghton Mifflin Harcourt Publishing Company.

Williams, Lisa. "The Kingfisher" and "Grackles" from *Woman Reading to the Sea* by Lisa Williams. Copyright © 2008 by Lisa Williams. Used by permission of W. W. Norton & Company, Inc.

Woloch, Cecilia. "For the Birds: A Charm of Goldfinches" used by kind permission of the author.

Wright, James. "You and I Saw Hawks Exchanging the Prey" from *Above the River: The Complete Poems* by James Wright, introduction by Donald Hall. Copyright

INDEX OF POETS AND POEMS

INDEX OF POETS AND POEMS

Index of Birds

. . . .

Numbers in italics refer to pages on which illustrations appear.

INDEX OF BIRDS